S0-BZS-840

Fifth Edition

Thinking About GIS

Geographic Information System Planning for Managers

Roger Tomlinson

Esri Press
REDLANDS | CALIFORNIA

Esri Press, 380 New York Street, Redlands, California 92373-8100

 Fifth edition 2013. Fourth edition 2011

Printed in the United States of America

17 16 15 14   2 3 4 5 6 7 8 9 10

Library of Congress Cataloging-in-Publication Data

Tomlinson, Roger F.

Thinking about GIS : geographic information system planning for managers / Roger Tomlinson.—Fifth edition.

pages cm

Includes bibliographical references and index.

ISBN 978-1-58948-348-4 (pbk. : alk. paper) 1. Geographic information systems. I. Title.

G70.212.T66 2013

658'.05—dc23   2013011693

Ask for Esri Press titles at your local bookstore or order by calling 800-447-9778, or shop online at esri.com/esripress. Outside the United States, contact your local Esri distributor or shop online at eurospanbookstore.com/esri.

Esri Press titles are distributed to the trade by the following:

In North America:
Ingram Publisher Services
Toll-free telephone: 800-648-3104
Toll-free fax: 800-838-1149
E-mail: customerservice@ingrampublisherservices.com

In the United Kingdom, Europe, Middle East and Africa, Asia, and Australia:
Eurospan Group
3 Henrietta Street
London WC2E 8LU
United Kingdom
Telephone: 44(0) 1767 604972
Fax: 44(0) 1767 601640
E-mail: eurospan@turpin-distribution.com

Dedicated to Mr. Leonard Hassall, geography teacher at Newmarket Grammar School, Suffolk, England, who turned my heart and life toward geography.

R. F. T.

Contents

Foreword

The author of the book you are holding has earned recognition and respect in the GIS world over the years. While Roger Tomlinson has rightly become known as the "father of GIS" as a result of his early work in using computers to model land inventories for the Canadian government in the early 1960s, I believe that his greatest contribution to the field is the rigorous method of GIS planning that is described in this book. For years he has advocated that one of the key ingredients to successful geographic information systems (GIS) is the use of a consistent planning methodology. He developed a methodology that has evolved over the years and can be easily adapted with the forward march of technology.

At the Esri International User Conference and other venues, Tomlinson's method is taught as part of a very popular "Planning and Managing a GIS" seminar. This seminar tends to attract two primary groups of people: The first group is composed of senior managers who oversee GIS and other information technologies in their organizations. The second group is composed of more technical managers responsible for the actual implementation of GIS and other information technologies. That these two markedly different groups come together year after year to glean Tomlinson's wisdom always strikes me as significant and as an excellent starting point for a book on GIS planning.

Roger Tomlinson wrote this book for those two kinds of managers, intending to bridge the communication gap between them. Senior executives in public- and private-sector organizations often have the general idea that GIS would be good for their organization, and they know how to get the resources allocated to make it happen. What they lack is enough understanding about the capabilities and unique constraints of geospatial data technologies to direct their technical managers (the second audience) and ask the right questions. Conversely, these line GIS managers tend to have a solid grasp of the technology and the unique characteristics of GIS but know much less about how the GIS must operate within the broader context of the organization itself. What they need is information that will allow them to answer the questions their bosses are going to ask. This book effectively and successfully serves the needs of both groups.

While these are the primary audiences, the book also has value for the student of GIS who wishes to learn how to do the middle manager's job. It is an invaluable source of tuition for students to understand what being a GIS manager in a large organization is all about and what they have to be able to do.

Successful GIS implementation depends on a well-thought-out and executed plan. If you follow the methodology presented by this book, you will be on the track to success. I hope that you find his work as informative and beneficial as have my colleagues and I.

Jack Dangermond
President, Esri

Acknowledgments

The errors are still mine. The methods described in this book have evolved over the years with help from many people. These include the associates of Tomlinson Associates Ltd. in Canada, the United States, and Australia. Many contributions were made by my colleague Larry Sugarbaker as we collaborated on the original "Planning and Managing a GIS" seminar together. It is impossible to thank Dave Peters too much. He has been my strong right arm in developing the current seminar series on which this book is based, and in particular the approach used in chapter 9 and the City of Rome case study. His research is the basis for the platform sizing and for calculating bandwidth requirements, particularly for the refined approach presented in appendix D. Generous input has been received from the staff of our clients worldwide and the staff of the corporations that eventually served our clients. They have been the real-world laboratory in which the ideas were tested. This book exists because Jack Dangermond thought it was a good idea to show the world what we put him through and because Brian Parr and Christian Harder turned the methodology into readable words. Amy Collins edited the fourth and fifth editions. She has brought a deep well of knowledge to this continuing work, which has allowed timely revision and completion of the text. This coupled with an eagle eye for detail has kept the project moving forward. We could not have done it without her. Peter Adams of Esri Press put this team together. His support has been patient and constant. None of this would have come about without the continuing work and friendship of my wife, Lila. Not only has she read every word and corrected most of my mistakes, but she still smiles. To these I owe my heartfelt thanks.

Roger Tomlinson

Introduction

A GIS project is as good as its plan.

If you have been charged with launching or implementing a geographic information system—a GIS—for your organization, this book is for you. Perhaps your professional field is one in which GIS has been used for many years, such as a local government, a transportation authority, or a forest management agency. Or perhaps it is one to which GIS has only recently been introduced, such as a corporate marketing firm, a political action group, or a farm. Either way, your organization is on the cusp of discovering the exciting implications of geographically enabled decision making.

The GIS you've been tasked with implementing could be intended to serve a single, specific purpose or to perform an ongoing function. Or it could be what's called an *enterprise GIS*, one designed to serve a wide range of purposes across many departments within your organization. You'll learn as your GIS evolves that a well-planned implementation can start out as a small project and grow, or scale up, into a full-blown enterprise system. An enterprise system that connects multiple government agencies is often referred to as *federated GIS*, which has a very broad scope that benefits entire communities.

Whatever the mission of your organization or the intended scope of the initial GIS implementation, the good news is that the fundamental principles behind planning for a successful GIS are essentially the same. These principles are based on the simple concept that you must think about your real purposes and decide what output, what information, you want from your GIS. All the rest depends on that.

This book details a practical method for planning a GIS that has been proven successful time and again during many years of use in real public- and private-sector organizations. It is a scalable approach—the methodology can be adapted to any size GIS, from a modest project to an enterprise-wide system—and one that can be reiterated as a GIS project grows over time.

Why plan?

So why should you plan? What's wrong with just buying some computers and GIS software, loading some data, and sort of "letting things happen"? Can't you simply adapt as things move along, tweak the system, learn as you go? In fact, doesn't all this advance thinking slow things down and create even more work? On the contrary, evidence shows that good GIS planning leads to GIS success, and absence of planning leads to failure.[1] Whether you are working with an existing system or creating a GIS from scratch, you must integrate sufficient planning into the development of your GIS; if you don't, chances are you'll end up with a system that doesn't meet your expectations.

Knowing what you want to get out of your GIS is absolutely crucial to your ultimate success. Too often, organizations decide they want a GIS because they've heard great things from their peers in other organizations, or they just

1 "Failure" meaning that after spending money on the system, you aren't getting what you want out of it, and this underperformance is evident to management. It could even result in the loss of someone's job.

don't want to get left behind technologically. So they invest considerable sums of money into technology, data, and personnel without knowing exactly what they need from the system. That's like packing for a vacation without knowing where you're going. You pack everything from your closet just in case, but it turns out that a sweater isn't needed in Fiji and you forgot the sunscreen. You've wasted time and energy, and worse yet, you're still not ready. When you try to develop a GIS without first seriously considering the real purpose, you could find yourself with the wrong (expensive) technology and unmet needs.

You must determine your organization's GIS needs from the outset of the planning process. GIS has many potential applications, so it's important to establish your specific requirements and objectives from the beginning. That way you will avoid the chaos that results from trying to create a system with no priorities or ends in mind. The methodology described in these pages will show you how to describe and prioritize what your organization needs from a GIS, so that you can plan a system that meets these requirements.

The key undertaking of managers, and those who plan on their behalf, is to understand their business and identify what would benefit that business. From GIS, the fundamental benefit comes in the form of what we call *information products.* An information product is data transformed into information particularly useful to you—for example, economic data analyzed in relation to a specific location—and delivered to you, via computer, often in the visual form of a map. If it's something that helps you do your work better, faster, more efficiently, then it's an information product.

Your GIS can quickly become a money pit if it's not creating useful products for the organization, ultimately jeopardizing the very existence of the GIS initiative and perhaps your own job. Conversely, a GIS can prove its worth and justify its existence if it manages to help streamline existing workflows and create useful information products. These are the ultimate benefits reaped by any successful information system.

Once you've identified the information products you seek, you can determine what data you need to make them. Then you can deal with the issues of tolerance to error and the conceptual database design on which efficiency will depend. From the type and amount of handling the data requires to make it usable (data requirements), you can define the system scope by specifying the capabilities you need from software (software functionality), and what your system requires in the way of support from the hardware and the network (hardware and network requirements). From these itemized necessities, you can develop accurate cost models to allow for clear and meaningful benefit-cost analysis. Having laid this groundwork, you can identify issues affecting implementation—institutional, legal, budgetary, staffing, risk, schedule, or timing issues—and look at how to mitigate them. The end result is an effective, efficient, and demonstrably beneficial GIS within the organization.

Implementation and maintenance can be expensive, but good planning will make your ongoing GIS efforts cost-effective in the long run. This book will teach you to evaluate the benefits of the system relative to its cost and how to make the case to management in a way that makes them advocates for your own success.

The entire planning process can take some time, and you may find that some of the steps can be minimized or eliminated in certain situations. But it is nonetheless important to think carefully about each step, to really get your head around the subject. You'll be glad you took the time.

GIS means change

Technology changes under us like the swell of a tide. To harvest the immense long-term benefits of GIS, you have to plan ahead in fast-paced times.

Rapid advancements in technology, both in software and hardware, continue to exert their strong effect on the GIS planning process. Improvements in software usability and advanced off-the-shelf GIS functionality mean faster, more sophisticated GIS development than ever before. The hardware that supports this grows ever more affordable—CPU seconds are approaching zero cost. The days when we designed system architecture around the limitations of software and hardware are over.

Now the driving determinants in system design are the location of human and data resources in the organization and the communication between them. Systems and communications distributed via the web are becoming increasingly important. Quite complex applications can be done on the server level now, and server virtualization reduces both cost and IT management complexity. Computer infrastructure requirements can be greatly reduced by taking advantage of cloud computing, where a cloud service provider sells computing resources on an as-needed basis. That's where the technology is going—follow it, and you won't be left unsupported.

Geospatial data also has become more accessible and plentiful, due in part to the increase in geographic measurement being driven by new technologies (GPS [Global Positioning System] receivers, lidar, etc.) and by real-time sensors capturing data to make it available as web services. Many standard and commonly used datasets are now readily available in digital form and at a much lower cost than even just a few years ago. A wide selection of worldwide basemaps can be easily downloaded from the web, often for free.

This relative abundance of affordable and reliable spatial data significantly widens the scope of potential GIS applications. Rapid prototyping and development tools such as Esri's ModelBuilder, Microsoft Visual Studio, and CASE (computer-aided software engineering) technology allow for quick exploration and testing of such applications. Map and model templates and code sharing are enormous time savers, allowing you to create information products more quickly than ever before. In other words, you can now explore a greater range of options as you work toward a customized solution for your business needs. The scalable nature of modern GIS means you can do targeted planning on selected business areas and build databases incrementally, growing them as needed.

These days, most GIS users handle spatial data within one of five development environments: In the first, the traditional stand-alone desktop information system, the user can conduct an integrated set of GIS functions on a wide variety of data types. In the second, the developer environment, software developers can combine a set of application-neutral, individual function components to create new applications. The third is the server environment. Here, a set of standardized GIS web services (e.g., mapping, data access, geocoding) support enterprise-wide applications and social media that allow multiple users to enter, view, manipulate, and analyze data from their offices or the field. The fourth is the online environment, with cloud-hosted integration and collaboration services, and the fifth is the mobile environment, which answers the proliferation of smartphones and handheld devices with mobile applications.

These five environments currently interoperate but are moving rapidly toward more unified models and interfaces. The more enterprise systems are in place, the more interoperability standards will be required. Integration is becoming essential, no longer a novelty.

Geographic information systems integrate seemingly disparate information quickly and visually, which facilitates communication, collaboration, and decision making. Through GIS, geography is actually becoming an organizing tool. In much the same manner as enterprise-wide financial systems converted the way organizations were managed in the 1960s through the 1980s, now geographic information systems are transforming the way organizations and government agencies manage their assets and serve their customers or citizens.

The focus has shifted from application-oriented architecture to a server-oriented one, making real-time geographic information available to anyone with access to the Internet. The results of GIS capabilities—quick complex analysis, maps showing statistical connections—used to be limited to the few. Now advances in server capabilities and the expanding use of cloud computing platforms means GIS technology is not only fully emergent on the web, but is quickly becoming standard on mobile devices as well, offering affordable and direct access to complex information products. The combination of GIS server technology and intuitive, easy-to-use web clients has opened up the GIS domain to everyone.

We've learned that individual geographic information systems themselves tend to evolve over time, but now we're discovering the value of GIS as a facilitator for a kind of organizational evolution. GIS means change—a new GIS implementation (into an organization that hasn't had GIS before) is a change agent. But once in place, GIS capabilities can be used to help an organization adjust to change, forecast changes ahead, and take advantage of the opportunities incumbent with change.

The ever-widening availability and continual advancements of technology have made it impossible to ignore that we live and work in a perpetual atmosphere of change. Planning is no longer a one-time event, but rather an ongoing process. More and more, organizations and public policy makers have come to identify GIS as important to their objectives. But to harvest the potential of GIS requires coordination, collaboration, and an enterprise view of GIS management. GIS planning itself is increasingly important if all these objectives are to be achieved.

How to use this book

Because the GIS planning stages are presented in the order in which they should be completed, the chapters of this book should be read consecutively. Once read through, however, the book can be treated like any other valuable reference book. You can use the detailed table of contents and the index to point you to a specific concept; dog-ear pages; write in the margins; take notes as you read. It is my hope this book will become a well-used tool as you work your way through the initial planning process, and as you continue to refine and expand your GIS project.

The supplemental DVD has a variety of helpful resources. You can view a recording of an entire eight-hour GIS planning seminar, or challenge yourself by completing exercises and quizzes that reinforce the material in the book. You can view examples of various documents produced in a real-world GIS planning scenario. Finally, you can increase your understanding of GIS concepts and terms by accessing a digital dictionary of GIS terms.

About this edition

The fifth edition of *Thinking About GIS: Geographic Information System Planning for Managers* incorporates the latest advances in software, hardware, server technology, and mobile solutions. Recent lessons learned from consulting practice have brought about new recommendations regarding the alignment of GIS capabilities with the enterprise management accountability framework, formation of teams, and best practices for effective communication and buy-in from upper management at large organizations. The in-depth case studies in chapters 6 and 9 have been revised to reflect current GIS practices. Updated "Focus" sections provide supplemental information about the methodology or current GIS technology. You will also find tips, definitions, and helpful resources throughout the text. While the book has been updated to make it more relevant to today's reader, the core message and proven methodology remain fundamentally unchanged. The ancillary materials on the accompanying DVD have also been updated with the latest available information. We hope you enjoy this new and improved edition.

Chapter 1

GIS: The whole picture

No GIS can be a success without the right people involved. A real-world GIS is actually a complex system of interrelated parts, and at the center of this system is a smart person or team who understands the whole.

GIS is a particularly horizontal technology in the sense that it has wide-ranging applications across the industrial and intellectual landscape. For this reason, it tends to resist simplistic definition. Yet the first thing we need is a common understanding of what we're talking about when we refer to GIS. A simple definition is not sufficient. In order to discuss GIS outside the context of any specific industry or application, we need a more flexible tool for elaborating: a model.

Figure 1.1 presents a holistic model of a functional geographic information system, which turns data, through analysis, into useful information. At the center, you can see that GIS stores spatial data, replete with its logically linked attribute information (from the left), in a GIS storage database, where analytical functions and programs are controlled interactively by a human operator to generate the needed information products (shown on the right).

Let's understand the GIS model better by examining its individual components. Spatial data is a term with special meaning in GIS. *Spatial data* is raw data distinguished by the presence of a geographic link. In other words, something about that piece of data is connected to a known place on the earth, a true geographic reference. The features you see on a map—roads, lakes, buildings—are ones commonly found in a GIS database as individual thematic layers. Most can be represented using a combination of points, lines, or polygons. Linked to these geographic features, and usually stored in a table format, is nonspatial information about them, data such as the name of the road, seasonal temperatures of the lake, owner of the building. These various characteristics applied to place are called *attributes* in GIS parlance, and, in fact, it is the range and depth of these attributes that make spatial data such a powerful tool in the hands of a dynamic, working GIS.

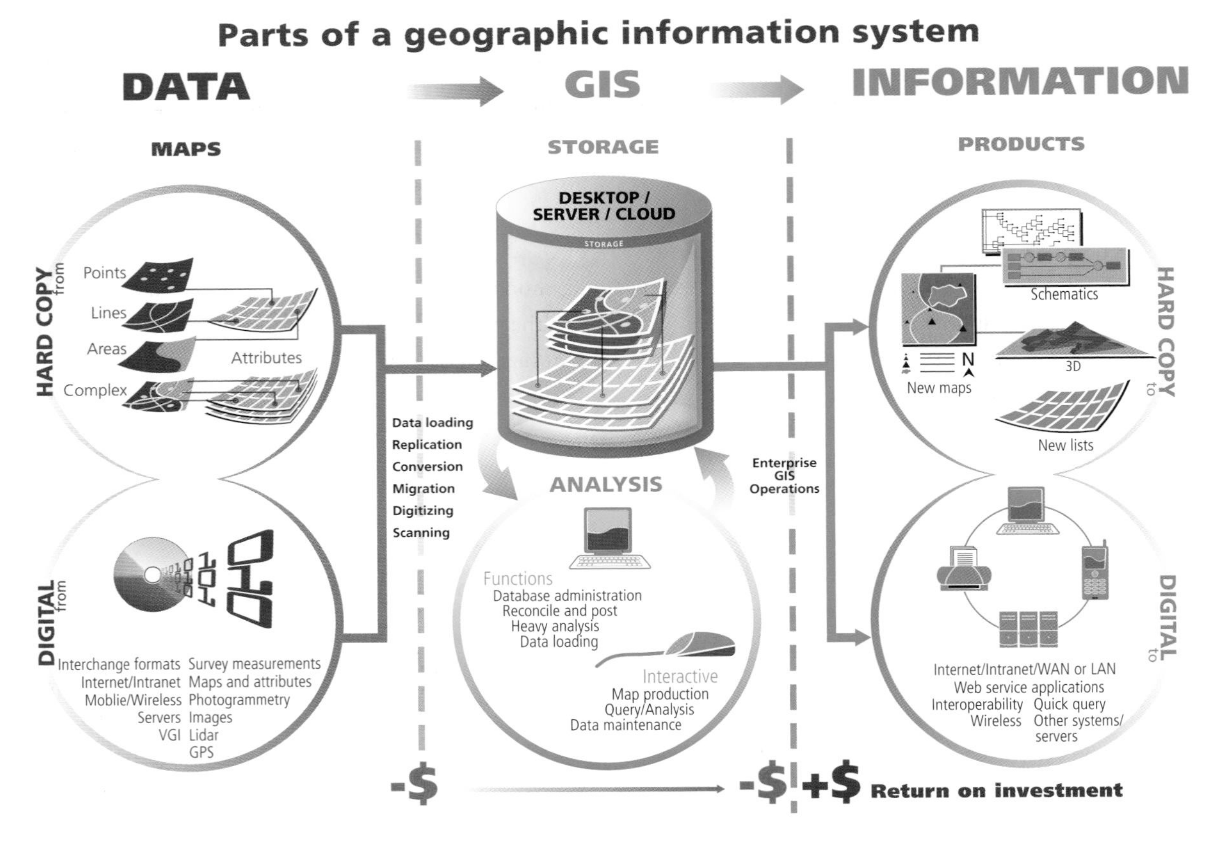

Figure 1.1 **Parts of a GIS.**

So where does this data come from? Not surprisingly, good old-fashioned paper maps and other hard-copy records still supply some of the important physical and human data needed for GIS. After all, printed paper maps have been the standard vehicle for conveying geographic information since the earliest recorded history; however, by scanning or digitizing the features drawn on our organization's paper maps, we mine this rich data source. And by establishing logical links to other digitized hard-copy records in our organization—tables, lists, documents—we further convert data for use in the GIS, doing this until we've digitized and linked all the relevant paper documents at our disposal. More and more, spatial data is available in digital form; you can buy it or acquire it via data-sharing arrangements or over the web.

Measuring and survey devices, including GPS receivers, photogrammetry images, and survey instruments generate troves of GIS-usable data, which can be shared quickly by means of the Internet and common interchange formats. All these sets of data with logical links—after being systematically integrated under the primary organizing key of geographic location—can be stored and managed as a unit, called a *database.*

Along with all its features and attributes, this linked unit of spatial data—the GIS database—resides in the GIS storage system, whether it be a desktop computer, server, or cloud service, where it is available for software functions such as analysis and mapmaking. The power of the computer is used to ask questions of the spatial data, to search through it, compare it, analyze it, and measure it. You use the GIS to do things that would be very laborious or even impossible to do in any other way. These GIS software functions are under the interactive control of the GIS operator, whose job is to create the needed information products, or may be executed under the control of a batch program for repetitive work.

Identifying the information products your organization needs is central to the GIS planning process, so we do it at the beginning. Information products come in many forms—new maps, new lists and tables, schematics, 3D visualizations, the results of interactive queries presented on-screen, as hard-copy maps and reports, or as transmittable digital information—but they are all intended to improve job performance. And when these end-product reports, which inform your choices, actually lead to better decision making, you know you've planned your GIS well. This is the harvest, the accomplishment that represents the ultimate success of GIS.

If you want to learn more about GIS capabilities and its myriad applications, several web resources and training opportunities are available. Start by visiting GIS Lounge (http://gislounge.com) or the Esri training site (http://training.esri.com).

Scope of GIS projects

Understanding the scope or range of operation of your project will help you develop an effective plan for GIS implementation. Is it a department-level application, a multidepartment enterprise, or a multiagency, federated system designed to benefit entire communities? The same guiding principles of GIS planning apply to all three scopes, regardless of subtle differences between project types (figure 1.2), but some of the planning steps may not be needed on small projects or department-level applications.

Organizations often end up testing the GIS waters with a single-purpose project carried out within a single department. The expected result is a project-specific output, such as information needed to make a decision. A site analysis to locate a new landfill is an example of this modest scope: a one-time effort that has an end date, with the project paying the acquisition cost and no long-term support expected.

The second level of GIS implementation is also within a contained scope, but without the limited time frame. With a department-level application, the objective is just as straightforward, but this time the need is ongoing: a department expects output from GIS to support at least one established business objective or function. For example, any time a change in land-use zoning is proposed, the city planning department must notify all property owners within three hundred feet of the property in question. The business objective is notification of all those owners; GIS supports this by generating the appropriate mailing lists. The GIS is located right there in the department responsible for the targeted workflow, and this department manages the system. For this reason, support from the departmental head is crucial; developing the GIS application depends on it. Funding to cover GIS staff, hardware, software, applications, and maintenance requires corporate approval.

Enterprise-wide systems are broader in scope, allowing employees to access and integrate GIS data across all departments of the organization. Here, fully aligned with the organization's mission, GIS takes a more versatile, active role. The objective is for GIS to boost an already established strategic direction, supporting the entire organization over the long haul. Advocacy from upper management is essential, as is long-term support from multiple departments. Enterprise GIS addresses the business needs of many or all departments, becoming a powerful tool inside the organization as a whole. For example, in a transportation company using multiple GIS applications and huge databases of geospatial data across all departments, GIS is a mission-critical element of the company's operating strategy. Corporate involvement is integral to ensuring data sharing among the many divisions. An enterprise-wide GIS allows the integration of this data with the business functions and processes.

Federated (aka community) systems are the broadest in scope of them all, serving multiple organizations (typically government agencies, and/or community service providers). In this type of project, benefits are often measured by overall community good that is achieved, rather than the success of one corporation. There is a centralized management of the project, but different organizations may be responsible for their own information products. For general discussion purposes, a federated GIS is often classified as a large enterprise undertaking.

The power of GIS can be leveraged the most at the enterprise and federated levels, where there is much to be gained from the adeptness of GIS to bring people and knowledge together: with consistent information available across the organization(s), decision makers get a clearer picture of reality; data is regularly updated; and more data is shared, reducing duplication of effort.

As GIS software continues to expand more and more into enterprise implementations, other trends lead industry observers to predict that the next step will be society-wide—GIS will become as much a part of our lives as computers are today. GIS servers are providing streamlined software architecture to enable multiuser access, simplifying the capability for serving maps and data on the web in a ready-to-use fashion.

The potential of GIS as integrator is only beginning to be tapped, but already GIS-based applications have brought modest changes into the daily lives of people, just as GIS has brought major change into the daily business of many organizations. As GIS-based applications become more widespread and available, who knows what societal changes will develop?

Interest in web services and in service-oriented architecture (SOA) is growing, and with it the public's access to knowledge previously available only to GIS specialists.

Departmental systems	Enterprise systems	Community/Federated systems
• GIS supports at least one important business function within a department	• GIS supports strategic business activities throughout the organization	• GIS supports the common goals of a community of enterprises
• Benefits often measured by productivity gains within the department	• Benefits often measured by overall success of the organization	• Benefits often measured by overall common community good
• Managed by the department responsible for the business activity	• Centralized support requirement (multiple departments)	• Centralized management of services benefiting multiple enterprises
• Periodic technology acquisitions	• System life cycle and procurement planning	• Common standards and service support agreements for resource sharing
• Corporate support important but not essential	• Corporate management support is essential	• Community support for central leadership is important

Figure 1.2 **Comparison of GIS project scopes.**

The who, what, when, where, why

Let's borrow a page from the reporter's notebook and set this story up via those famous "W"s of the newsroom: *who, what, when, where,* and *why?*

Who should plan a GIS? You, the GIS manager (in a large organization you might be known as the GIO, or geographic information officer), must take the lead role in the planning process, but you should never go it alone. You need the senior-level decision makers on board with you, advised and informed throughout the process. Failure to keep these budget keepers apprised can lead to reduced or eliminated funding. To ensure their support, keep them actively engaged in the planning process and educated about the work. You rely on them also to tell you what information products are going to be needed at their level. Include in the planning process those who will be directly using the system as well. If they aren't involved, you'll probably fail to meet their real needs.

A note on consultants: If you decide to hire GIS consultants, have them lead you through the planning steps—never hire a consultant to do the planning for you. You and your colleagues need to do the planning—it's your GIS system, your job, and your reputation at stake. You'll be making decisions as a team throughout the planning process. If you do use a consultant to help you, a GIS team within the organization should still carry out the work, under the consultant's guidance.

If you decide to hire a consultant, make sure he or she is committed to creating a custom GIS plan for your organization. You don't want someone who simply recycles the same reports and design specifications. The best way to make sure you're hiring a consultant who is worth the cost is to ask for references and check them personally.

In Canada there is a saying: "Consultants disappear like the snow in springtime." The point is, at the end of the process you will be left with the system to implement. If you haven't been totally involved in planning and writing the implementation strategy, you might be in for a very difficult and painful time.

What to plan? GIS is a complex system of interconnected parts. So it will come as no surprise that you must consider six different major components in any GIS plan: information products, data, software, hardware, procedures, and people.

Information products: Information products are what you want (need) from the GIS. This desired output may take the form of maps, reports, graphs, lists, or any combination thereof. Mission critical: identify these products with sufficient clarity early in the planning process.

Data: By knowing what information products you want, you can identify the data required to make them and plan for the acquisition of the needed data. What can you get that already exists? What can you create from existing sources? What levels of map accuracy and scale will you require? And don't forget data format, a factor related to the next component, software. Sometimes data format alone drives the software decision, as in the case of the city municipal government planning a data-sharing arrangement with a county GIS department that already uses a particular software.

Software: Software programs provide the functions needed to perform analysis and create the information products you want. Sometimes customized software sits on top of the main GIS software package. Updates need to be planned to keep the versions current. There are also support and operating system issues related to software.

Hardware: GIS is demanding of hardware. You must take an unflinching look at your organization's computational resources and upgrade accordingly to support your GIS. Typically, a few powerful workstations support the heavy lifting and geoprocessing, though in larger systems this is increasingly handled by GIS servers on a network or in the cloud. Simple computers ("thin clients") on the network provide the user access for database query and display purposes. A robust internal network and wide bandwidth ("fat pipe") to the Internet are also required to facilitate file sharing, data acquisition, and reporting.

Procedures: As an important component of GIS planning, *procedures* refers to the way people do their jobs and the changes they will have to make to do their jobs using your new GIS system. You need a migration plan to facilitate this transition from the old way to the new, plus you need to address how the existing ("legacy") systems will coexist (or not) with the GIS.

People: GIS requires the right people, people with skills of geographic analysis, problem solving, and adequate technical capability. Above all, the people must be able to think. Will you need to hire or are the right ones already on staff? How will you hire, train, and keep the staff with the specialized skills it takes to build or use your system? Over time, staffing will be your single biggest cost.

When to plan? You plan from the beginning, and the planning process continues after the GIS is installed. GIS planning is no longer a one-time event, especially in an enterprise or community system. Successful GIS projects attract positive attention, which means that before long people will identify other things that they'd like to see from the GIS. This will move you to revisit one or the other of the planning stages. Armed now with the empirical experience of a functioning GIS, each new iteration of the planning process becomes better calibrated to the real world.

Furthermore, you plan in a particular order, each step illuminating where to go with the next. This only stands to reason. How could you plan to acquire the data for a map, for example, unless you'd already identified that map as part of an information product you need?

Where to plan? GIS planning, to be most effective, must be carried out in the business world, not from the vacuum of your office. Because GIS has the potential to

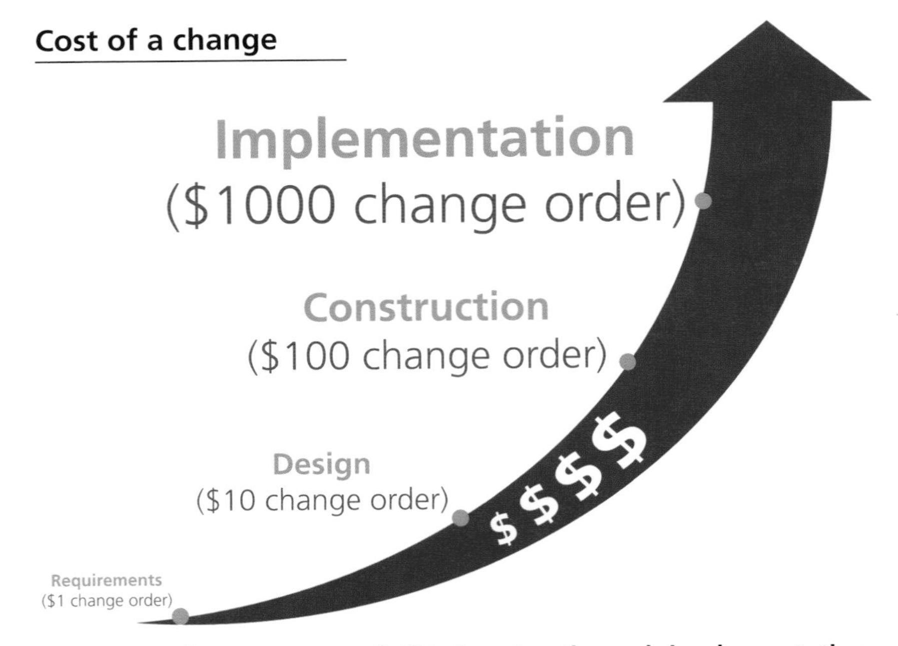

Figure 1.3 **The cost curve of GIS planning through implementation.**

create common connections between disparate things, it is inherently a horizontal technology that can touch literally every person in an organization if the company's leaders want it to (and many do). The thorough and diligent GIS planner must meet people in the organization to learn what information they need and how they need to get it. Only by taking the time to witness people doing their work can the GIS planner ever truly understand the business's processes. And how can a planner create anything of use to anyone without this knowledge? For an enterprise GIS this may involve travel to multiple organizations and sites. Large projects have even greater need of consistent and effective communication, so be prepared to *go.*

Why plan GIS? Good planning leads to success and poor planning leads to failure. This is true whether you are starting from scratch or building from an existing GIS. It seems obvious on the surface, but time after time GIS projects fail because of poor planning. Like any complex information system, GIS implementation and maintenance are costly. Every component of the system—the data, the software, the hardware, the staff—cost an organization dearly (figure1.3). But everything exacts a lesser toll if it is considered beforehand and selected carefully. An enterprise undertaking without meticulous planning could be disastrous.

Would you rather spend a dollar at the beginning on planning or ten thousand later to make up for not planning? The point is, even spending a lot of money is no assurance of getting the right thing. Besides, planning can be interesting: you meet a lot of people, learn a lot about your organization, and may end up knowing more than the CEO about how the place really runs. Planning is especially rewarding toward the end when you do the benefit-cost analysis and see how much money your organization will be saving over time—and all because you led the implementation of a well-planned GIS.

Chapter 2

Overview of the method

Like a good road map, an overview of the method lets you know where you are going.

The planning methodology introduced in this book shows you the steps of GIS planning—how to assess what your requirements are and which system will meet those needs—and how to implement the GIS in your organization once your plan is approved.

This nine-stage GIS planning methodology evolved from years of experience in planning large and small implementations in public- and private-sector companies. The size and nature of your organization will determine which of the component stages are most relevant to your situation. A full enterprise-wide implementation almost certainly requires you to undertake all the stages in full, while for a smaller project you'll be completing some steps quickly or even skipping a few. Regardless of the size of the undertaking, all situations are unique; you will need to understand all of the steps in the process before adapting the methodology to suit your circumstances.

The nine-stage GIS planning methodology

Stage 1: Consider the strategic purpose
Stage 2: Build the foundation
Stage 3: Conduct a technology seminar
Stage 4: Describe the information products
Stage 5: Consider the data design
Stage 6: Choose a logical database model
Stage 7: Determine system requirements
Stage 8: Consider benefit-cost, migration, and risk analysis
Stage 9: Plan the implementation

The following chapters of this book detail each of the nine stages. Let's take a quick tour of the method.

Stage 1: Consider the strategic purpose (chapter 3)
Start by considering the strategic purpose of the organization within which the system will be developed. What are its goals, objectives, and mandates, and what sort of management accountability framework is in place to measure and evoke success?

This stage of planning ensures that the GIS planning process and final system fit within the organizational context and truly support the strategic objectives of the organization. This stage also allows you to assess how information created by the GIS will affect the business strategy of the organization.

Stage 2: Build the foundation (chapter 4)
GIS planning should not be taken lightly. Forget about actually implementing a GIS for the moment. Just planning a GIS takes a commitment of resources and people. Before you begin, you need to know that your organization understands the distinction between planning and implementing and that it is prepared to provide enough resources for the planning project.

Making the case means understanding what needs to be done and what it will take to get it done. The result of this stage is a planning proposal that makes that case and explicitly seeks approval and funding to launch the formal planning process.

Commitment to the planning process is essential to a successful GIS implementation, especially in municipal government agencies and other bureaucratic public-sector organizations. The planning proposal helps secure this political commitment. This is the moment to introduce the GIS planning process to the highest level executives of your organization and to arrange to keep them fully informed of your progress. If you receive approval for your planning project and a commitment of resources at this point, your chances of having a successful GIS are high.

Once approval is secured, your in-house GIS teams should be assembled. These teams are especially important in large organizations—planning a GIS is not a one-person job!

Stage 3: Conduct a technology seminar (chapter 5)
Once your planning project is approved, you can activate the in-house GIS planning team to begin its most important endeavor: identifying exactly what the organization needs from a GIS.

Defining the specific GIS requirements is the primary task of the planning process. You must meet with the customers or clients of the GIS (those who will use the system or its output) to begin gathering specifics about the organization's needs from the user's perspective. A highly effective method of soliciting input is to hold one or more in-house technology seminars.

The technology seminar is an ideal opportunity for you to explain to key personnel the nature of GIS, its potential benefits, the scope of the planning process itself, and to solicit their input. It may also be necessary to introduce the basic methods of geographic analysis. The technology seminar is the place where initial identification of information products begins. By involving stakeholders at this early stage, you help to ensure subsequent participation in the planning work ahead.

Stage 4: Describe the information products (chapter 6)
Knowing what you want to get out of your GIS is the key to a successful implementation. And what you will want will come in the form of information products: maps, lists, charts, reports, whatever you need to inform your decision making and streamline workflows.

This stage must be carefully undertaken. You will talk to the users about what their job involves and what information they need to perform their tasks. Ultimately you need to determine things like how each information product should be made and how frequently, what data is required to make it, how much error can be tolerated, and the benefits of the new information produced. You will help each person declaring a specific need for such information from the GIS to write an *information product description (IPD).*

This stage should result in a document that includes a description of all the information products that can be reasonably foreseen, together with details of the input data (the *master input data list* or *MIDL*) and functions required to create these products.

Stage 5: Consider the data design (chapter 7)

In GIS, data is a major factor because spatial data is relatively complicated. In this stage you will review spatial data fundamentals so you are better equipped to plan the data design that will best model the information products you require. In addition to spatial data characteristics, you must consider data sources, data and technology standards, and conversion and interoperability guidelines. Doing so will help you refine your information product descriptions, determine your output data requirements, and be conversant in factors that affect data design strategies.

Stage 6: Choose a logical database model (chapter 8)

A logical database model describes those parts of the real world that concern your organization. The database may be simple or complex, but it must fit together in a logical manner so that you can easily retrieve the data you need and efficiently carry out the analysis tasks required.

Several options are available for your system's database design. You will review the advantages and disadvantages of each approach at this stage, so you can choose a model that will best represent your data and its unique relationships, attributes, and behaviors.

Stage 7: Determine system requirements (chapter 9)

Here, you will envisage the system design in its entirety by examining as a whole what you will require of a system: the GIS workstations, software, communications bandwidth, and core capacity, and their location (place or cloud). This is the first time in the planning process that you will examine software and hardware products.

You will review the information product descriptions and the master input data list (discussed in chapter 6) in order to summarize and classify the functions needed to make these products. This will enable you to inform vendors of what you require in the way of software functionality. You will consider issues of interface design, effective communications (particularly in distributed systems), and platform sizing in order to determine the appropriate hardware, software, and network configurations to meet your needs. You will submit your recommendations to management with a preliminary design document.

Stage 8: Consider benefit-cost, migration, and risk analysis (chapter 10)

Following conceptual system design, you will need to work out the best way to actually implement the system you have designed. This is where you will begin preparing for how the system will be taken from the planning stage to actual implementation.

As part of that preparation, you will need to conduct a benefit-cost analysis to make your business case for the system. To convince management to fund the GIS implementation, you will probably be called upon to show how various risk factors weigh in, such as migration from the old system to the new.

Stage 9: Plan the implementation (chapter 11)

Until now, the focus of the planning methodology has been on what you need to put in place to meet your requirements. The focus at this stage switches to how to put the system in place—acquisition and implementation planning. Now you will address such issues as staffing and training, institutional interactions, legal matters, security, legacy hardware and software, the implementation schedule, and how to manage change. The final report that results from this last stage of the methodology will

FOCUS

Let each step inform the next

- If you know what information products you need, you can determine what data should go into your GIS.
- If you can determine what data should go into your system, you can also determine what needs to be done to those data to produce your information products.
- If you know what you want to do to your data, you can determine what functions your system needs to be able to perform and begin to design an appropriate technological solution.

contain your implementation strategy and benefit-cost analysis. The report can be used both to secure funding for your system and as a guide for the actual implementation of the system.

The final report equips you with all the information you need to implement a successful GIS. It will become your GIS planning book to help you through the implementation process.

Developing the final report should be the result of a process of communication between the GIS team and management so that no part of the report comes as a surprise to anyone. The report should contain a review of the organization's strategic business objectives, the information requirements study, details of the recommended data design, recommendations for implementation, time-planning issues, and funding alternatives.

See components of a real-world final report on the supplemental DVD in the back of the book.

The purpose of this GIS planning methodology—and my intention with this book—is to guide you through these stages in your thinking. Use it to give senior executives the context for the questions they must ask about GIS in their organization; let it inform you as a planner or new GIS manager how to answer those questions.

Test your GIS planning skills. Take the "Gauge your knowledge of GIS planning" quiz on the supplemental DVD now.

Chapter 3

Consider the strategic purpose

Strategic purpose (or a government mandate) is the guiding light. The system that gets implemented must be aligned with the purpose of the organization as a whole.

It all starts with the organization. To develop an effective GIS, the GIS planner must have a clear understanding of what the agency or company does, its working plan to do it, and how GIS can help accomplish the mission. Organizations adopt GIS on the assumption that it will make their work easier and cheaper to do, or that it will be better for the customer or constituency. Fostering any or all of those GIS benefits begins with understanding how the organization works. A successful GIS is one that is aligned with the purpose of the organization as a whole, thereby helping provide what the organization needs to stay true to that purpose.

How do you find out what your organization needs? As the GIS planner, first you examine the strategic business plan; most organizations have one. It states the goals as they are envisioned plus the actions required to meet them. A strategic business plan may include the following components:

Mission statement or mandate: Describes the purpose of the organization. This may be a concise, pithy statement of one or two sentences, or it may be a lengthier statement that includes the history of the organization, its values or guiding principles, its vision for the future, or a list of high-level goals. For example: The mandate of Parks Canada is to protect and present nationally significant examples of Canada's natural and cultural heritage and to foster public understanding, appreciation, and enjoyment in ways that ensure the ecological and commemorative integrity of these parklands for present and future generations.

Program directions: The current direction or strategy of the organization's efforts toward fulfilling its mission. In other words, if you were to add up everything the different programs are doing, the sum of these efforts should achieve the overall organizational goals. For example, program directions in a tech company might include:

Implement a common data management system; promote compatibility and interoperability among data producing departments; institute a cloud-based system for software distribution and data storage.

Business model: The specific manner in which organizational mission and program directions are to be achieved. The business model also provides insights into business sustainability—how revenue will be available for business costs over time. A business model might include the following elements:

- Management accountability framework: A performance management tool that codifies specific, department-level, measureable goals, often set with deadlines, as well as the proposed means to obtain said goals, thereby carrying out the mission of the organization. An effective performance management tool should allow executives to assess the performance of individuals and departments without bias. Just like it sounds, these goals are meant to keep management and employees accountable: if x, y, and z have been achieved (by the deadline, if applicable), performance is assessed positively. With successful goal achievement, you can imagine there would likely be a positive effect on yearly performance evaluations, promotions, raises, or bonuses, if applicable. It may be as simple as a list of expectations, or it may be called something else, but the intent is the same: to support and assess performance, particularly of managers or department heads.
- Employee development and support: The plan for providing employee training and staff development.
- Public interaction: How the public or your constituency participates in developing and updating the strategic plan. Such involvement can be direct or through indirect methods such as surveys or focus groups.

See examples of mission statements on the DVD in the back of the book.

To see an example of a management accountability framework at the federal level, go to the Treasury Board of Canada Secretariat website (`http://www.tbs-sct.gc.ca/maf-crg`**).**

Every bit of this information, in any form you can find it, will contribute to your overall understanding of the organization's strategic direction and purpose, which will inform your GIS planning. Be creative and thorough in discovering what the organization is all about. If it does not have a strategic plan, examine its mandates and responsibilities to get a sense of organizational purposes and objectives. The planner's job is easier, however, if the organization has an established strategic plan; if senior management's objectives match those of the strategic plan; and if there is commitment throughout the organization to achieve its goals. It's also important to note whether technology is prominent in the picture; for sustained support of GIS development, technology must be recognized as an important tool in achieving success.

The organization's strategic plan provides the context within which you will do a GIS benefit-cost analysis (chapter 10) as well as your assessment of what additional information would help the organization accomplish its goals and objectives. Understanding what the organization does and its vision for the future allows the GIS manager to design valuable information products, ones that further those objectives. Without considering the organization's strategic purpose, you risk wasting time on planning that is peripheral to its needs.

If the organization does not have a strategic business plan, it may take some work to find out what the real business is and thereby what the real needs are. But even a relatively detailed business plan tells only part of the story. You need to uncover the organization's secrets, its objectives, how its workflows really operate, what makes it tick. You need to know all of this to plan for an effective GIS.

To really understand the business, you must also analyze the mandates and responsibilities of each functional division that will be involved with GIS. Undertaking this analysis requires active engagement of the stakeholders. Go into individual departments and seek answers to these questions:

- How do the individuals who make decisions currently do it?
- What do they need to know to perform their tasks?
- What information products are appropriate for these tasks?

Imagine the following scenario: You go into a government department of forestry to talk to the staff and begin by asking a forester, "Can you please tell me what you do here?"

"My job is to manage the harvesting of timber in the best interests of the people of the state," she replies.

OK, that's how she sees her job. Next you ask, "What do you need to know to do that?"

She talks around the subject at length, listing many of the datasets she uses, before she finally gets to something tangible: she really needs to know which trees to cut down, when to cut them down, and how to cut them down. This simple statement contains the crux of the information you need to know, but you had to ask probing questions and be a good listener to filter out all the rest. Now that you know what she really does, you can start to identify some information products she will need on her desk, information products that will actually tell her which trees to cut down, when, and how, based on the methodology she already uses.

It will take several conversations like this to get a handle on the information that individuals need to carry out their jobs, fulfill their mandates and responsibilities, or accomplish the objectives of the strategic business plan.

Make sure you obtain a clear understanding of the responsibilities and current workflows of departments and individuals within your organization who may benefit from GIS.

Photo courtesy of Shutterstock; ©Pressmaster.

Asking these questions should help those you interview focus on the specifics you're after:

- What are your job responsibilities?
- What do you have to achieve?
- What do you have to produce?
- What do you need to know to carry out your responsibilities?
- What information can GIS produce and put on your desk that will provide what you need to know and help you monitor or keep track of your responsibilities?

In this way, the strategic direction of the organization will reveal itself and, with the answers to these questions from a range of individuals, you will be able to envisage the scope of the information they need to succeed. Answers given by management may differ significantly from those of the rank-and-file, so you will "blend" the responses in order to get a clear picture of the information required from GIS.

Asking the same questions in different departments will help delineate workflows within departments and their interaction with workflows in other departments. Ultimately, a comprehensive overview of the organization's processes will emerge. With this insight, the GIS planner can begin to examine the following:

- What data is available now that can be used to create the information products needed?
- Where is this data?
- What new data is required?
- What data-handling functions are necessary to turn the available data into the required information products?

When you know what functions are needed and what data these functions must operate on, you have enough information to define the technological requirements (e.g., hardware and software).

The link you establish between strategic objectives, information, and data gives you an audit trail. You identify how information fits into the organization and the benefits of creating new information. This is the start of a benefit-cost analysis for GIS: what is the benefit of creating information products for the organization, and what is the cost of producing those products?

Once the information products are available, they will help the organization to fulfill its mandates and responsibilities, meet its objectives, and progress in the direction intended. Even better, the new information products allow the organization to change its strategic direction in response to new opportunities or to identify new markets. This moment, when an organization's leaders finally grasp the full strategic implications of their GIS, is a sweet moment indeed for the GIS planner.

Chapter 4

Build the foundation

Since the GIS planning process will take time and resources, you need to obtain corporate approval at the front end. Because the planning process requires more than the work of a single individual, assembling strong teams secures the foundation.

After you have a clear vision of your organization's strategic objectives, you are ready for "foundation planning." Planning for a GIS usually requires a serious commitment of time and money. It can take from six months to a year to look at the overall needs of a large organization, during which time you'll be depending on its personnel and funding. That's why you need specific permission to plan, along with management's commitment to provide the resources. You ask for both within a planning proposal document, which you will write and submit as soon as possible. The planning proposal document sets forth a process for planning that is tied into the strategic plan of the organization in particular, the management accountability framework, which encourages those who are to be involved in the planning process, particularly managers, to feel seriously invested in the project. GIS planning should be considered as high a priority as implementation, and should be rewarded as such. Once the planning proposal is approved, you will assemble your GIS teams, solidifying a foundation from which to launch the planning phase.

A planning proposal is a useful tool for making the case to management for the resources to complete your GIS planning project. This proposal should explain exactly what's involved: the stages and costs of planning, in time and money, including the use of in-house staff resources. It will outline the steps described in this book, tailored exclusively to your organization.

Develop this proposal early to get the organization behind you. Make sure that management signs off on your proposal before going further. With management support beginning at this stage, your chances of a successful GIS implementation are high. Approval and commitment of resources are essential, and they need to come from the top—from senior management. You might have to start discussions with middle management and work through the chain of command, but eventually you will have to sell the idea to the higher-ups, all the way to the person on top. It can't be said enough: it is *strongly* recommended that you get the CEO's buy-in from the outset.

Your proposal can make a convincing case to top management by pointing out that both the success rate and the benefits of GIS increase significantly with better planning. In fact, it's the planning study you are proposing that will give the GIS planning team the opportunity to identify what the organization needs most from a GIS: the specific information products that could bring such improved efficiency into its day-to-day work.

It is also worth mentioning in the proposal that planning is more important than ever, despite positive changes in the economics of computing that foster the illusion you could just dive in. In the early days of GIS, when the massive hardware installations required to run the systems could cost more than a million dollars, organizations looking at GIS could easily justify spending significant time and resources on planning. With that kind of money at stake, they had to make sure they were going about things wisely. Ten percent of the total system cost was an entirely reasonable amount to allocate for planning. But in today's computing environment, with adequate hardware and software available for thousands rather than millions of dollars, 10 percent of the system budget would amount to only a few thousand dollars, not an adequate amount for in-depth planning. The proportion of funds needed in the planning stage relative to the overall budget has shifted. Five to seven percent of the first year's overall GIS budget is typical. (Read more about overall cost in chapters 9 and 10.)

These days, the ultimate cost of GIS implementation will depend heavily on data rather than hardware or software. Therefore, while your planning project proposal will include cost estimates for hardware and software as well as information on anticipated staffing needs, it will have to show that the most significant investments will be in data acquisition and development, not in hardware and software. The data costs for your GIS may surpass the hardware and software investments, once you factor in things like measuring equipment, labor, conversion

FOCUS

GIS and management accountability frameworks

Many organizations worldwide have some form of a management accountability framework. As mentioned in the previous chapter, a management accountability framework is a key performance management tool used to support the accountability of deputy heads and to improve management practices across departments and agencies. The management accountability framework establishes agreed-upon outcomes for enterprise activities and establishes measures to determine the resultant degree of success. It is recommended that GIS information product benefits are expressed in two ways: first, the contribution that the information product will make to one or more successful outcomes identified in the management accountability framework; second, the fiscal benefit to the line-item budget (internal order code) which will be impacted by the information product.

Under the direction of the chief executive officer (CEO), the chief administrative officer (CAO) is responsible for establishing GIS governance and implementing monitoring of GIS costs and benefits in line with the management accountability framework. You will read more about benefit-cost analysis in chapter 10.

costs, maintenance, and the licensing of commercial data.

As your investments in data grow, so too will the time and resource requirements of ongoing maintenance. So despite the perennial lowering of costs for equipment, the need for thorough and thoughtful GIS planning is as important today as it ever was.

The planning proposal

Let's look at a couple of successful planning proposals. The first one was developed by an in-house GIS advocate working at a national park; the second by a GIS consultant, laying out a planning proposal for an Australian state government effort. At Jasper National Park in Canada, the in-house planning team submitted a document called a *terms of reference* to the park service senior management with the purpose of seeking funding for the GIS planning process. The organization of the proposal is shown in figure 4.1.

Jasper National Park
Terms of reference

Work outline

1.0 Project description
- 1.1 Background
- 1.2 Objectives of the project
- 1.3 Project deliverables

2.0 User needs analysis
- 2.1 Situational assessment
- 2.2 Client base
- 2.3 Business requirements and information products
- 2.4 Data requirements
- 2.5 Technology requirements

3.0 Software/hardware assessment

4.0 Database requirements

5.0 Implementation plan

6.0 Schedule of payment

Contract conditions
- Parks Canada responsibility
- Timing and duration
- Proposal guidelines

Appendix 1 Documents for review

Appendix 2 Ecosystem database inventory

Appendix 3 Client list
1. Jasper National Park clients
2. Corporate clients
3. Tri-council research
4. Agency clients
5. Private industry clients and general public
6. Information management

List of bidders for GIS user needs

Figure 4.1 **Outline of an in-house planning proposal.**

Figure 4.2 illustrates the sections of a successful proposal submitted by Tomlinson Associates Ltd. to the state government of Victoria, Australia, for involvement in a large-scale GIS planning project. Bear in mind that this is a document produced by consultants aiming to get work assisting the GIS planning process.

Strategic framework for GIS development

Introduction
Project team
Outline of the proposed methodology
- Staff seminar
- Information product definition
- Benefits
- Data requirements
- Error tolerance analysis
- Information priority
- Functional requirement
- Benefit–cost analysis
- Implementation planning
- Request for tender design (technical contents)
- Strategy report

Project deliverables
Work organization
Procurement process (optional)
Project timing
Involvement of government staff
Administrative and cost notes
Cost assumptions
Cost summary
- GIS planning project
- Cost breakdown—by fiscal year
- Cost breakdown—by primary year
- Optional cost—electoral office procurement process

Appendix 1: Background and experience of Tomlinson Associates Ltd.
Appendix 2: Curriculum vitae of key personnel

Figure 4.2 **Contents of a GIS consultant's planning proposal.**

A note about hiring consultants: they can help, but you must be the one who drives the process, works with it, writes reports, and makes trade-off decisions. A consultant should not entirely do your GIS planning—you must assume that responsibility. When the consultant leaves, you will be left with the system to implement. If you think a consultant is right for your GIS planning project, be sure to consider the consultant's fees in the planning proposal, along with the other anticipated costs of the planning phase (which are separate from the implementation costs, discussed in chapter 10).

These examples give you the general idea about what constitutes a planning proposal. You can adapt them to the specific needs of your own organization. Your proposal should be detailed and deal with all aspects of the GIS planning process. Although there are no set guidelines for creating GIS planning project proposals, it is a good idea to deal with each phase of the planning methodology in a distinctive section (as in figure 4.3). Within each section, include potential staffing needs and staff time commitment. Figure 4.3 illustrates the basic sections of a proposal using parts of the GIS planning methodology.

Once you present your proposal, it is imperative that it be reviewed and approved and the resources committed. You have not asked for money just for hiring a consultant; you have requested a commitment of in-house time and staff resources. Since the GIS planning process will use these resources, you need to get approval and commitment before you take the next step. Incentivizing management by connecting the planning outcomes to an accountability framework (think performance reviews, bonuses, promotions) will ensure that the planning process will be carried out. Make sure that management signs off on your proposal to plan for a GIS, and ask the CEO to write a memo expressing the importance of GIS to the organization, ensuring his or her belief in the project, and requesting the support of directors or upper-level managers.

Assemble the teams

After you receive the go-ahead, you can prepare to evaluate your organization's requirements in more detail, and for that you need teamwork. Now that you have commitment to the planning process, you'll need to confirm who should be involved in it and brief all the participants on their roles and responsibilities. It is crucial that an organizational framework is put into place before planning commences. This is considered a foundation planning task, because if you don't have a strong base of effective leadership, management support, dedicated personnel, and clear task descriptions, you are building your GIS plan on shaky ground—not a good starting point. In addition to writing the planning proposal, foundation planning includes assembling a strong GIS planning team composed of people who are committed to exploring the full potential of GIS for the organization. It also

Project description
- Background
- Objectives
- Deliverables

Requirements study
- Preparation
- Needs assessment
- System scope

Conceptual design
- Database design
- Technology design
- Hardware
- Software

Implementation planning
- Benefit–cost calculation
- Risk analysis

Figure 4.3 **Stages of the GIS planning.**

means organizing a management committee who is going to deal with administrative and funding issues, thereby easing your path to success. As you build the planning teams, keep an eye turned forward, to the implementation stage. Anticipate the personnel that will be required for the implementation stage, and avoid surprises. Figure 4.4 outlines the recommended teams that will need to be assembled for the GIS planning project (an implementation team, which may have different members than the planning team, is discussed in chapter 11). Keep in mind that while the chart identifies titles, what's really important is that the skills typically associated with those titles are represented. One person may be able to fill the requirements of two chart entries, or one entry may be best filled by two employees. Each organization is unique, so you may choose to modify the paradigm presented here. Also keep in mind this isn't intended to be a complete staffing chart; rather, it identifies the key players that will make the planning and implementation phases go smoothly. These players will need to communicate their individual staffing needs, particularly when it is time to implement the plan.

The planning team

Your planning project needs to be conducted by an inhouse GIS planning team. Ideally, this team should be helmed by the GIS manager (you, perhaps), and composed of two or three other people who have the desired skill set. One person should have participated in at least one previous GIS planning study and be fully conversant with the methods and techniques employed; it might be a GIS planning specialist, or consultant. You also want someone who qualifies as a GIS technical expert—someone who can identify opportunities as well as limitations within the most current software (this might be the GIS manager or it may be another individual). Finally, you should have a business representative, someone from inside the organization who will use one of the information products created by the GIS. In a larger organization, you may have several business representatives, one from each department, requesting information from the GIS. These individuals will assist by clarifying the needs of each department. Note that some GIS projects—public websites, for instance—are designed mainly for outside clients. In this case it might be helpful to gather user feedback in some capacity at the planning stage—focus groups or customer surveys, perhaps—but the planning team is best comprised of internal personnel.

After the planning team has been established, you can meet with department heads. The memo you obtained earlier from the CEO or director of your organization stressing the importance of the GIS planning process should have requested the support of all department heads. As you meet with the department heads, you will probably find that they want to know the following: What is the role of the department in the planning process? What time commitments will be necessary? Who will be needed? In response, you can answer their questions, making it clear that some of their personnel will be involved in the planning and may need to spend several days in the process. This must be stated up front so that staff are given time away from their normal duties to help with the planning with full support of their manager.

If all you're undertaking is a single-purpose project within your own department, you may only need to visit one department manager, your own, to get the green light. Even so, it is important to get support, and to stress that the planning will take some time and that you will not be able to do as many other things during this period.

For a project of any cross-departmental scope, you should definitely visit all the department heads separately to secure their support. Find out what is needed for them to reach a level of comfort with your GIS endeavors.

An important note: keep in mind, right now you are assembling a team for planning only. Later on you will need a dedicated implementation team, and most likely an entire staff who will be asked to spend part, if not all, of their time on the GIS side of business. As you move through the planning, if you find employees or managers who are especially fired up about the GIS project, try to find a role for them once it's time to implement the plan.

A strong team guiding the planning process with consistency in approach and method inspires confidence and cooperation. You'll need both for the all-important task ahead.

The enterprise planning team

Common sense says a large project needs a larger planning team. Modern enterprise-wide GIS planning can involve various activities at multiple sites; thus, enterprise projects require a formal management structure. A successful planning strategy for multisite organizations has been to employ a GIO (geographic information officer) whose sole responsibility is to manage the planning and execution of GIS projects. Then training is provided to a group of site facilitators who, under the direction of the GIO, return to the sites and complete the GIS planning with a site team. The overall enterprise plan is the combined and coordinated output from each site plan.

The management committee

To be effective, a GIS manager needs the structured support of higher-ups. Especially important for enterprise projects, a *management committee,* preferably appointed by the CEO and comprised of the CAO (chief

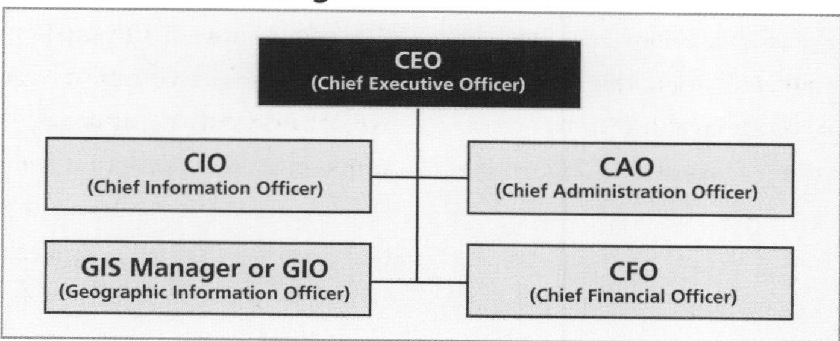

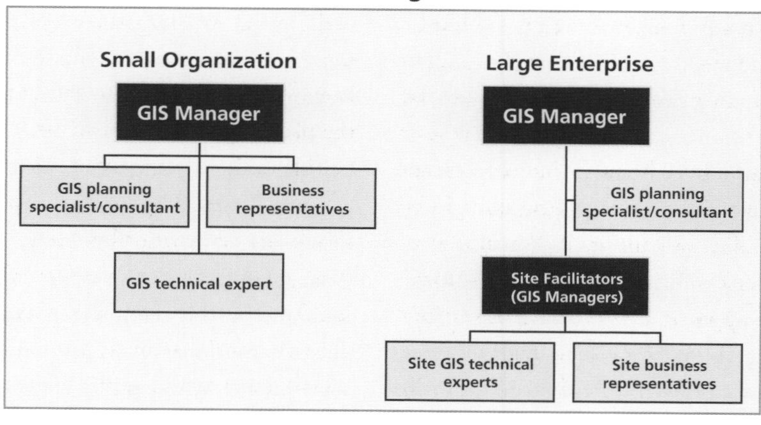

Figure 4.4 **Recommended planning teams.**

administrative officer) or equivalent, CIO (chief information officer), GIS manager (or geographic information officer, GIO), and the CFO (chief financial officer), is the champion of the planning process. For an enterprise project it is essential that the CEO be fully involved and supportive. It is not recommended that enterprise-wide GIS planning be undertaken without CEO support and involvement.

The management committee (sometimes called the steering committee) formally manages the project scope, approves the schedule and budget, ensures timely availability of needed resources, resolves major policy and personnel issues, provides guidance and direction for key business strategies, and expects regular updates on the planning status. The management committee is formed at the planning stage but continues to function through implementation and beyond. It is the committee's job to oversee the integration of the GIS plan into the organization's strategic business plan, and to make sure the role of GIS in the strategic plan is maintained.

The management committee should coordinate scheduling to avoid timing delays. Several actions must be coordinated in time if the planning process is to be efficient. For example, in large organizations that will employ site facilitators, these facilitators must receive training. But delays in the start of the planning process result in too large a gap between training and planning, and subsequent skill loss in the facilitators. Such delays are frequently caused by funding obstacles, which should be on the agenda of the management committee.

One of the management committee's key tasks is to handle "back-fill" personnel—the new hires for vacant (or soon-to-be-vacant) positions. If a GIO or GIS manager position is unfilled, it must be advertised and filled before planning commences. It is no easy task finding the right people and going through the necessary hiring process; this can be especially time-consuming in government agencies. It is desirable to have a minimum three-week overlap between incumbents leaving their positions and the new back-fill employee, to allow the new person to gain competence in the working practices of the organization. The same goes for open positions that will affect the planning and implementation stages, such as GIS analysts and systems administration staff. Indeed, putting the right people into the right place at the right time is one of the most critical tasks of the management committee.

Read about GIS staff, job descriptions, and training in appendix A.

The crucial role of GIS manager

A well-planned GIS project needs a strong leader. The most important person in the equation is the GIS manager, or the GIO. The larger the organization, the greater the opportunity for obstacles, delays, and slowdowns. Bureaucracy and red tape can cause funding delays which can impede start-up at sites; lack of direction can leave site facilitators floundering; organization-wide disruption (labor disputes, for example) can halt progress. It is imperative that large projects have firm leadership to ensure strong momentum and keep sites on-target. A strong leader will not only deal with funding issues, staffing problems, and technical issues, but will stay energized and focused to keep the troops fired up. For instance, many enterprise projects have stalled due to lack of early mentoring or monitoring of the site facilitators. That is the job of the GIS manager, or the GIO, so visiting site facilitators—early and regularly—must be a priority.

While a dialed-in management committee will be a huge asset, it is the GIS manager who is responsible for keeping the project on the active radar screen for all the higher-ups. It is imperative that the GIS manager is able to communicate directly and effectively with upper management. Working closely *with,* not just taking orders *from,* the management committee is very important. The committee is designed to oversee *your* project, so you must give direction, and allow them time to consider requests and needs.

In short, the GIS manager must not only have a firm grasp of GIS technology and capabilities, but also be a meticulous organizer, a strong leader, and an effective communicator. Management and communication workshops are available for people who may need to gain strength in these areas.

A note about the organizational structure

GIS planning may result in temporary changes to the organizational hierarchy. Typically organizations are hierarchical, with the straightforward tree-and-branches form of reporting. As the scope of the GIS project begins to involve several parts of the organization, matrixed responsibilities and reporting are often adopted. Experience has shown that in enterprise-wide GIS planning, reporting should transition to project-based management responsibilities: specifically, the site facilitators at different sites (departments) should be responsible to and report directly to the organization's GIO for the duration of the GIS planning phase. This has proven most effective for adequate project control and focus on project objective and on time and budget envelopes.

Given that at the end of the planning phase the management committee approves the project and its implementation, the recommendations should be made part of the organization's business plan. At that point, the department heads are responsible for the GIS implementation within their departments. The site facilitators revert to their departmental GIS managerial roles within the departments, reporting to the department heads in the normal hierarchical structure.

When the planning proposal has been approved and a strong team assembled, foundation planning is complete, and you are ready to assess the GIS requirements of your organization.

FOCUS

Plan ahead for the time commitment

These guidelines regarding resource requirements are based on the experiences of the author in planning a number of GIS implementations at a variety of regional and municipal organizations:

- From start to finish, the GIS planning process in a typical region, government, department, or small municipality takes four to eight months, longer for complex or large organizations.
- The aggregate total of work hours amounts to six or seven person-months, more time for complex or large organizations.
- The leader of the in-house GIS planning team (the GIS manager) is committed for 70 percent of his or her working time (and would probably opt for more if other duties didn't intrude). For large enterprise projects the leader will need 80 to 100 percent of his or her time.
- Departmental staff involved will spend four to eight days in the development of each information product description (more about this in the next two chapters).
- All levels of staff will be affected in some way and are thus indirectly involved with the study.

Chapter 5

Conduct a technology seminar

Think of the technology seminar as a "town-hall" meeting between the GIS planning team and the potential users of GIS in the organization.

Having studied the organization's strategic plan, you have the official word on the goals of the work it does; now you're going to find out how the organization actually works toward those goals and how GIS can help those workflows move more smoothly to their objectives. You're going to hold a big meeting—a kind of group study—with all the potential users to find out what people need from GIS to do their jobs better.

You have already assembled your in-house GIS team and secured the support of department heads for their staff to be involved, so now you can bring everyone together for a technology seminar. Think of this as a kind of town-hall meeting, wherein the in-house team shares the vision of GIS and reviews the planning process, while everyone affected has a chance to voice what they want from GIS. The deliverable for this meeting will be the initial list of information products, divided by department.

This is a training event to raise awareness of GIS and explain the roles of those involved. Depending on the size of the organization, you can host one or more technology seminars—whatever it takes for your colleagues to understand the planning process and the fundamental concepts behind GIS. One approach is to run an event over two or more days, in order to accommodate all the people who expect information from the GIS. In a large organization, you may have thirty or more participating; in a smaller one, maybe a dozen or less. For multisite organizations, you may need to coordinate seminars on each site, or you may take advantage of video conference technology.

Seminar components and tips

Before going into more detail, let's take a look at what it is about a technology seminar that moves the planning forward. You will see from the following description of one that a seminar supports the planning process by eliciting an itemized first listing of the output desired from a GIS and by gaining support from everyone by educating them about GIS. Of course, such events vary according to the nature of the organization, but basically the agenda for any GIS technology seminar should include the following:

- Welcome statement
- Describing a GIS
- Defining GIS terminology
- Explaining GIS functions
- Explaining the planning process—steps and responsibilities
- Eliciting preliminary identification of desired information products
- Explaining business workflow improvement opportunities

A welcome statement is self-explanatory, but if the CEO can be persuaded to say a few words it will help greatly with motivation and seriousness. Whether it's a recorded message or a brief stop in person, ask your CEO if he or she can take the time to get this project started off on the right foot and make its importance to the organization clear. A few minutes is all it takes.

Successful planning requires effective communication among those involved. Spend some time during the first day of the seminar explaining the nature of GIS, defining the basic terminology and describing the functions that systems can perform. This will help everyone in the organization establish a shared vision of GIS and a common language in which to articulate their requirements.

Participants want to know when and how they will be involved in planning and what is expected of them, so start by saying that this seminar begins their participation in the planning process. This is their first forum, and in it they will define the deliverables; by identifying the information products they need from GIS, they will be starting the planning process. Tell them you will open the discussion to audience participation shortly, but first you want to spend some time defining GIS and outlining the stages of the planning process. Explain how each planning step logically follows one upon the other, building on the information gathered in the one before. Emphasize that the first step—identifying the output that participants require from GIS—is the most important because it sets the course for planning. And this is why their involvement is so crucial: it's up to them to declare what they want out of the system (or at least what they think they want).

Then open it up for discussion and begin to capture what the participants say. Use a whiteboard and assign someone to take good notes. Ask people directly: what's the job you do and what information do you need on your desk? They will probably think about it and say they don't know what they need. Keep prompting them and eventually they will give you some ideas.

A forester with a timber inventory to manage might say she could use a better harvesting map. An urban engineer tracking sewer repairs might say he needs an accurate inventory of sewer hookups, one that's updated automatically whenever completed work orders are submitted back to engineering.

Even after you explain that in GIS we call this kind of output—anything that meets your specific workflow needs—an *information product,* some people will get hung up in limiting their thinking to maps, under the assumption that GIS is a mapping system only. So be sure to mention early on that sometimes the best GIS output comes in the form of lists derived from spatial analysis, such as a record of all the customers in a trade area. One result of GIS spatial analysis of all the properties in a flood zone could be the series of addresses of people to be contacted in an emergency, another example of new information coming in the form of a list. While such lists can be subsequently mapped, the real value of these information products comes from the GIS process of spatial analysis, not from its mapmaking function.

Because you're involved in a session of collaborative brainstorming, the first ideas will trigger others and before you know it, most people in the group will have described the workflow story in their own context. Write what you're hearing on the board, making a list of the information products identified. (Try to get consensus in the naming of each information product.) You can expect to collect ten, twenty, even fifty information product ideas by the end of the technology seminar. Most of the products should be meant to facilitate current workflow, but also include potential products that participants anticipate needing as their department advances—a "wish list" as opposed to a "needs list," perhaps. At this point, it doesn't matter which information products take precedence over others—some may never even get built—you simply want to start from the largest universe of possibilities you can muster. You can count on the process itself to lead to an emerging sense of priorities, as well as a great deal of cooperation and cross-fertilization of ideas.

FOCUS

Purpose of the technology seminar

The purpose of the seminar is to do the following:

- Introduce GIS to the participants (if necessary)
- Introduce the planning process to the participants
- Explain to participants the reason why the work is being done
- Make clear the nature of the contribution required of participants and how their efforts will improve the chance for success overall
- Introduce GIS terminology and methods that will be used throughout the planning and final implementation
- Afford participants the opportunity to assess their work needs and identify the information that would help them do their job better and more efficiently
- Compose the list of information products needed, by name and by department, including the name of the people who need the products to do their work (not their bosses)
- Include potential products for future phases of the business workflow, in addition to products currently needed

The technology seminar is the first face-to-face contact between the GIS planning team and the staff involved in the planning study, so an underlying objective of the seminar is to establish a good working relationship between the two groups. Toward this end, members of the planning team should demonstrate that they possess the following:

- Competence in GIS and, preferably, in the business of the company
- Experience in related GIS planning efforts elsewhere
- Ability to elicit the participants' views on information needs and management practices without pushing their own views
- Ability to aid the organization in clarifying and describing its own requirements

For now, all you need is the title of the proposed information product, a few words describing its intended use, the scale of any output maps required, and the name of the person who wants it. The person's name is very important; you may need them for more details later on. It also serves as an indication that someone wants the information product so much that they're willing to support it themselves. The person set to actually use the product is the one who really knows why it's needed, and he should be willing to associate his name with it. Make it a rule: no name, no information product.

One more thing: it is not a good idea to let the head of the department design all of the information products. An authority figure's viewpoint acts as a filter, and at this early stage we want the ideas coming in unfiltered. Along the same lines, in a large enterprise, one strong department (or even one strong site) may take an aggressive leadership role in the GIS planning process. This may create a risk that the set of information products may be unbalanced in favor of the strong host rather than the enterprise as a whole. Balance should be addressed in the seminars, but ultimately it is the responsibility of the management committee, especially the CEO. Management needs to ensure that adequate information products are produced to support the decision making that underpins the strategic business objectives of the organization.

Conducting a technology seminar in person (or through video conference) is preferable to brainstorming via e-mail. Getting good minds together in one room, talking face-to-face, at the same point in time, breeds greater creativity and energy.

Set the stage

GIS means change, and whenever you introduce change into organizations there will be internal debate and usually some resistance. Building trust and alliances with colleagues who will be affected by your efforts is an important dimension of achieving success. Enabling everyone to get prepared and stay informed is considerate and fosters trust, so you should distribute a set of well-written (and carefully proofread) documents to all participants prior to the seminar. Include in the package any relevant memoranda authorizing the study, the agenda of the first meeting, and the timetable for the study. Also provide some introductory GIS information that people can study before coming to the meeting. You could photocopy articles or sections of textbooks (with the publisher's permission) or even purchase copies in bulk of a certain book if you find one you like. (See "Further reading" on page 257 for a list of books appropriate for this purpose.)

You'll discover at the meeting that some people will have read your documents from front to back, others not at all, and most will have skimmed them with varying degrees of interest. Look for the people who really have taken to the GIS introductory materials; these are people who, like you, have recognized something inherently interesting about GIS and can be recruited as allies in your GIS campaign. These focused individuals may be good candidates for the GIS implementation team when it's time to launch the project.

Limit the number of staff invited to a single seminar to twenty-five or thirty people organization-wide, with each manager making the selection for his or her department. Smaller GIS efforts in smaller agencies obviously require smaller teams. Among the required attendees are the senior administrative officer of the organization and the heads of all departments involved. Their attendance alone, even if they are figureheads, sends a signal of senior management's commitment to the effort, which should stimulate interest in all invitees to attend. Especially encourage experienced staff members to attend; they bring intimate knowledge of their own department's processes. It may be useful to invite representatives of other organizations with GIS experience, such as those in neighboring states, regions, or provinces. But these should be carefully selected. The focus of the meeting is on your own organization. Particularly, include the people involved with any GIS efforts already established in your organization.

Arrange to hold your technology seminar away from the office—better for brainstorming. Local hotels, universities, and conference centers typically make for safe, easy-to-find locations with appropriate meeting facilities. The "no cell phone" rule is good practice. The ideal location should offer a large room with comfortable chairs and ancillary rooms nearby to allow the group to break up into smaller working units. The main room should have several flip charts or whiteboards (or an overhead projector and screen), ample marking pens, computer projection capabilities if desired, and one or two large tables. The smaller rooms will require flip charts or whiteboards as well. If off-site employees are participating, you will need computers, Internet access, and a video conference subscription.

Plan the program

The typical seminar takes place over two or three days. The first part of the meeting—or the first day—should include the following agenda items, which may vary in sequence somewhat:

- Welcome and statement of commitment to the project by the senior administrative officer of the organization, or, preferably, the CEO if available.
- Overview of current GIS status in the organization, including current GIS procurement or GIS activities, if any.
- Introduction to GIS by the team leader to establish basic GIS concepts and common terminology. This might include a definition of GIS, the parts of a GIS, and the functions of a GIS. This alone could take up a full day (or more, depending on the level of former knowledge of the participants).
- Explanation of the needs assessment process, with the team leader emphasizing that the contribution of participants is central to the work. A key goal of the meeting is for participants to describe the information products they need.
- Brief overview and examples of information product descriptions (IPDs).

Expect many questions and issues surrounding the descriptions of information products. Be prepared to explain the generic GIS functions in detail (see the "Lexicon" section on page 239) and stress the difference between data (the raw elements of information) and information products (those raw elements transformed into information useful in doing work). Show how interim information products will evolve, and how those used in one area might find application in other areas. Encourage people to look for these cross-department applications.

Assess information needs

On the second day (or after the earlier discussions if you have only one day), begin the initial assessment of information needs. This amounts to brainstorming, so encourage open-mindedness because, though the result you want is short and concise, the way to get it may be long and roundabout. At this stage in planning, all you're after is a brief identification of the information products needed: descriptive title, map scale, name of individual who needs it, and scope (other people or departments who might use it). The key to success in this step is to get participants to think about their overall information needs, freely and creatively as well as realistically, so that ultimately they come up with something that really works for them. This may be something they may not have thought of before, or even something that changes their workflow altogether. New ideas can be fruitful, so allow participants to spend some time on assessing their information needs. Don't let participants be discouraged by the amount of complex analysis they think it will take to create their ideal information products. Tell them that GIS is capable of extremely complex analysis, and if this product is deemed a priority there's no reason it can't become a reality. (Assessing the complexity comes with the next step, where you further describe the information products and outline the steps required to create them.)

Finally, you want to end up with a piece of paper listing a logical name for each information product, attached to the name of the person who needs the product on his or her desk to perform his or her work. One team member should act as reporter and fill in an "information products needed" form as each new idea is brought into the discussion.

Start the assessment of information needs by asking one department head to state broadly the responsibilities of the department. Write a concise version of this on the flip chart or whiteboard. Then ask other members of the department about their responsibilities. Ask them to identify the kinds of tasks they are responsible for, the types of decisions they have to make, the need for information at their workplace, and the conditions they regularly monitor in order to do their jobs. Capture the essence of this information on the whiteboard.

Now go back and start over, this time asking these same staff members to identify a single information product they could see as useful. Discuss the product only until you have a good idea of what it contains, making sure that it is an information product and not a dataset. (Remember, a dataset is simply a grouping of related data, while an information product is the result of one or more datasets turned into information in a form particularly useful in doing your work.) Decide on a brief descriptive title and write this on the board. Add the scope of the proposed product, that is, who else in the organization will require it. If it's a map-based information product, make sure you note the scale. Identify the name of the individual who needs the product on her desk to do her job. That person must be prepared to define it more clearly in the next stage of planning. If nobody is willing to "own" the proposed idea, drop it from consideration. This will prevent your list from swelling with mere idle thoughts and half-baked ideas. Identify a second information product from the same department and continue in this fashion until that department is out of ideas.

Now move on to another department. Repeat the process until there is at least one information product identified for each department. A great deal of cross-fertilization of ideas, corporate-wide understanding, and potential for cooperating are frequent outcomes of the seminar. Better ways of doing business can be recognized; opportunities for revised workflows based on GIS can be identified. This can be noted on the whiteboard for future examination. These benefits come when the group is kept together listening to the requests of other departments. If time becomes a constraint, and at the risk of losing some cross-fertilization, you may decide to split into separate rooms to allow each department to work further on its initial list. At the end of the day, bring the whole group together to clarify any problems and to collate a list of the information products needed. This allows you to check for duplication.

Figure 5.1 is an actual list of information products, organized by department, that was developed at an internal technology seminar for a city government. You can see that the product titles alone, in most cases, explain what they're about without more detailed information.

Information products for a city

Engineering and public works	Housing and property
Public works program map and list	Legal surveys index query
Sanitary sewer analysis	Select area plan production
Road needs map and list	Legal surveys drafting (CAD)
Pedestrian system needs	Dwelling unit analysis
Annual sewer needs map and list	Social housing acquisition analysis
Basemap flooding situation	Housing protection analysis
Sidewalk ramping needs	Housing potential map
Minor hard services needs	Residential loans and complaints map and list
Detailed engineering design (CAD*)	Existing housing and demographics map and list
Storm sewer analysis	City property map list
Maintenance analysis map and lists	Development activity map and list
Route optimization map and lists	Social housing map and list
Sidewalk snow removal analysis	Amenity map for selected area
Sidewalk route analysis	Waiting list analysis
Snow removal scheduling	Architect/landscape design drawing (CAD)
Lane map and public notification lists	Architect/landscape technical drawing (CAD)
Fleet tracking	Interior design drawings (CAD)
Waste receptacles/litter analysis	Streetscape design drawings (CAD)
Complaints analysis	
Park site maintenance analysis	
Stop signal location analysis	
Traffic/parking changes	
Parking needs analysis	**Fire department**
Traffic counts	Fire suppression water supply map
*Acronym for computer-aided design	Emergency response building floor plans
	Planning department site plans
	Fire emergency demographic analysis
	Emergency route selection
City clerk's office	**Recreation and culture**
Ward profile analysis	Recreation site analysis and design
Citywide election data map and list	Recreational/cultural site location analysis
Ward/poll/voter election data map and list	Recreational/cultural opportunity analysis
Economic development	**Legal department**
Development projects map and list	Lane and street index map and list

Figure 5.1 **List of information products, organized by department, developed at a seminar.**

(continued on next page)

Information products for a city (continued)

Planning and development	Planning and development (continued)
Site planning existing conditions	Employment analysis map and list
Permit parking program map and list	Accident rate analysis maps and lists
Development agreement analysis	Spot speed survey map and list
Zoning analysis	Development site impact analysis map and list
Cash-in-lieu-of parking map and list	Pedestrian flow map and list
Street and lane closure analysis	Intersection operation analysis map and list
Development information map and list	Employment/FSI capacity map and list
Circulation address labels and lists	Office and retail vacancy rates map and list
Reservation of special-needs housing locations	Summary of development projects map and list
Registered special-needs housing location map	Commercial activity map and list
Property location map and application cross-reference	Capital works project status list
Area monitoring map and list	Public land facilities list
Citywide monitoring map and list	Inspection status by district map
Official plan designations map	Cumulative development activity
Area monitoring map and list	Inspector performance activity map and list
Vacant land assessment	
Census tract monitoring map and list	**Police department**
Spatial market analysis	
Neighborhood monitoring map and list	Special-event planning
Transportation flow density map	Emergency route response map and list
3D simulation of urban section	Recent occurrence analysis map and lists
Perspective views of urban section	Police response property map and lists
Ward monitoring map and list	Crime occurrence—best suspect system
User-defined area monitoring map and list	
Overlay mapping system	

Figure 5.1 (continued)

List of information products, organized by department, developed at a seminar.

Rank the benefits

With this initial list of information products in hand, you should scan the list with an eye toward measurable benefits. Recalling the strategic business plan, you should ask how the information products identified in the technology seminar would benefit the organization as a whole. Now consider the more specific management accountability framework. How do the information products fit into it, and which budgets will each product benefit?

Review the list of information products with the GIS team and management representatives and check off the ones most important in terms of the value of the benefits they could bring. Try to rank them from most to least important.

This preliminary, ranked list of the most beneficial information products will help in the next stage of planning. You'll be selecting the most important or mission-critical ones from the list, and for those that are technically achievable, you will draft fully detailed descriptions. (You will find a more detailed account of these information product descriptions in the next chapter.) This is how your initial list spawns your next to-do list: the first round of information products to be developed.

When you are planning an enterprise-wide system with multiple sites, you may have trouble with timeliness: certain sites may take too long prioritizing the initial list of potential information products. This is usually caused by an "excess of democracy" and is cured by firm decision making on the part of the site facilitator, supported by the GIS manager. Intense project oversight is needed to keep multiple sites on task. Earning the reputation of a micromanager is better than being known for helming a stalled GIS project.

Go with the workflow

During both the technology seminar and while you're ranking the information products according to how beneficial each could be, you are delving into how the organization actually does its work, getting a sense of how various tasks flow from one to the other. This can be a very productive endeavor, with people thinking about their organization's workflows and the information products that could facilitate them. GIS planners need to understand these workflows because the GIS tools are often required to operate within a larger workflow.

In this business context, workflows are models of complex business processes used to gain more efficient operations within the organization. Typical complex business processes include things like land-use development approvals, building loan approvals, permitting, conservation land planning, power outage response, service delivery, and distributed facilities management.

When facilitating such a business workflow process requires more than one information product, you create these related and interdependent information products using one GIS application. An application is two or more information products produced from the same software and linked by a business workflow model, as in figure 5.2.

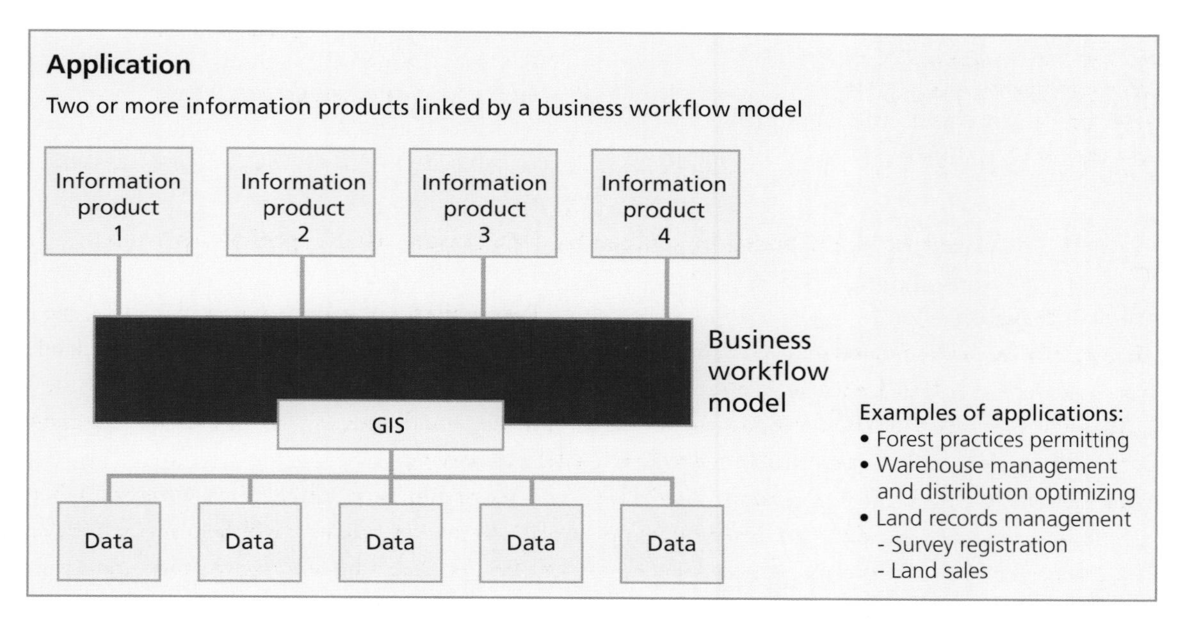

Figure 5.2 **A GIS application turns data into useful information.**

For example, imagine you work within a city's planning department. On a regular basis, people come into your office requesting building permits. This request triggers a well-established business workflow model that involves locating the property, determining if there are any building restrictions on the lot, and notifying all persons who own parcels within 200 feet of the permit location. Using GIS capabilities, you could develop an application program that locates the subdivision within which the lot falls (we could call this information product #1); displays a map of the subdivision, which shows any building restrictions like utility easements and building setbacks (information product #2); and finally, selects all properties within 200 feet of the permit location and exports their ownership records to a mailing list (information product #3). In this case, the application creates three information products that reproduce an existing workflow associated with approving a building permit.

Applications compose the centerpiece of GIS and the culmination of the planning process. If you ever get the chance, ask a GIS manager what applications his or her system is running. If they are savvy planners, their eyes will gleam as they tell you about these complex applications, the information they produce, and how they are streamlining existing workflows. You might get the same glimmer during the technology seminar when people start thinking about their workflows with an eye toward improving them. A typical scenario evolves when you

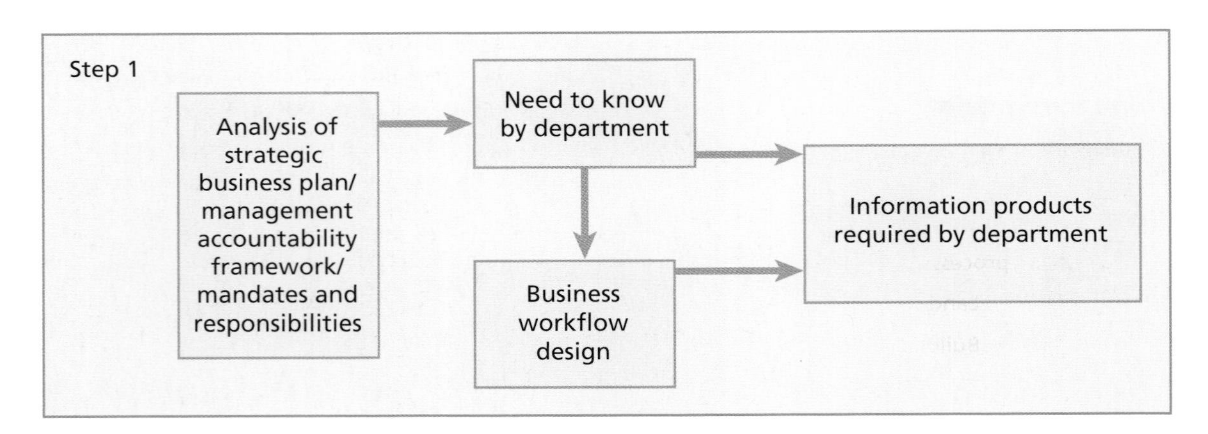

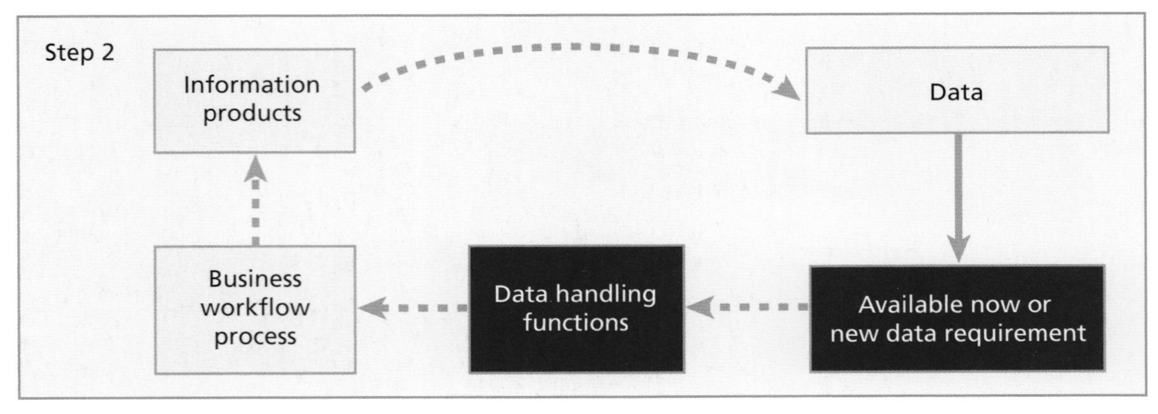

Figure 5.3 **Study logic diagram.**

begin to press them for what information products they want: people realize they need to take a hard look at their business processes at the same time. If they are having trouble determining information products, suggest that they define the actual workflow process and see what information products come out of that.

GIS is an enabling technology that allows people to think about changing their workflow (figure 5.3). So this is where you ask, "Do you want to redesign a workflow in any part of your organization?" And if the answer to that question is yes, you need to think about who's going to do it. If it's necessary to contract out the work, you must take into account how long it will take because it has to be completed before the GIS can be integrated.

What's the business workflow design and is there an opportunity to rework how the organization does business to gain efficiency or to capitalize on a new opportunity? Be sure to ask these questions and discuss the issue because this may be where a very high percentage of the benefit from GIS comes into an organization. Big gains result from the reduction of time or the number of steps in a process and from having a centralized database instead of replicated databases all over the place.

You must consider business workflows in your planning. In step 2 of figure 5.3, note that data and data handling move through the business workflow process. As you enter the next phase of planning, describing the information products and the data that comprise them (chapters 6 and 7), you will see why it's so important to keep this in mind.

FOCUS

Modeling workflow processes

Complex business processes designed to achieve more efficient operations are called workflows. In a workflow, typically the output of one process forms the input of a subsequent process. Examples of these complex business processes include the following:

- Land-use development approvals
- Building loan approvals
- Permitting
- Conservation land planning
- Power outage response
- Service delivery
- Distributed facilities maintenance

Typical tools used to describe workflow include:

- Visio Enterprise
- Python scripting
- Workflow Analyzer
- Enterprise Process Center (EPC)
- BPwin
- iGrafx FlowCharter

Chapter 6

Describe the information products

Know what you want to get out of it.

The technology seminar provided you with an initial list of information products and names of the people who requested them. This is your starting point. The whole unfiltered list frames the scope of expectations that exist in the organization at this time. The next step is to take the most urgent and important of these information products, as defined by your ranking, and start looking at them in detail. At this point, with your help, the people requesting them will begin to develop thorough descriptions of information products—specifications that will allow the actual product to become a working reality.

For the GIS to create an information product for you, you must first envisage it and write down descriptions of everything about it. These information product descriptions (IPDs) are the building blocks of the planning process.

In creating these IPDs, you are specifying for the first time the output your GIS must be able to generate. Once you know that, you can detail the prerequisite input on the master input data list (MIDL). After that, in chapter 9, you will use specifications derived from the IPDs and MIDL to configure the best system design to support your GIS, which you will recommend to upper management. These IPDs are the tools you'll leverage to gain approval for spending money on GIS hardware, software, and data. IPDs are very important, and it's at this stage in the planning process that you will create them by doing the following:

- Clarify the information products that need to be produced by the system
- Establish what data is needed to create the information products
- Identify the system functions that will be used to create the information products
- Assess the benefit to the organization of having each information product
- Create a master input data list and list functions needed to input source data (MIDL)

- Prioritize information products and data-acquisition needs

If this seems like hard work, you are being astute and realistic. This is the mental heavy lifting necessary to create well-defined specifications for the information products that you're ultimately going to build. "Pay me now or pay me later" warns the craftsman's adage. In some ways, because it is such a creative endeavor, this is also the most interesting part of the process. After this crucial step, what remains of planning your GIS will fall into place systematically.

The individual components of an IPD

Each information product description is composed of one or more forms, like the ones in the case study starting on page 56, but each descriptive component in the IPD does not necessarily require a separate form. For example, any electronic documents (text documents, scanned images, or videos) can go on the product's document retrieval form. Pop-up windows on maps are associated with map features, and should be noted on the map form, but the detailed contents of the pop-up window are best described on other forms—the list form if

FOCUS

IPDs: The building blocks of GIS planning

An IPD defines what is required to make the information product and what is expected of it. In particular, what you need from your information product in the way of maps, lists, and scanned documents especially helps in clarifying the output required from the system. Determining these output requirements in turn clarifies what must go into the system, the data input. To be useful, each information product description should include all of the following components that pertain to it:

- **Title** or name of the information product.
- **Name** of the department and name of the person who needs it.
- **Synopsis,** a narrative summary providing an overview of the information product in layperson's terms. Importantly, it should address how the information product meets specific goals as outlined in organization's strategic business plan or management accountability framework. One or two paragraphs is usually sufficient.
- **Map output requirements,** which consist of either a map sketch or an actual example from another source, including a legend. This can be 3D output. Make sure to include any pop-up information you want included in your map, as well as labels, symbology, graphic elements, schematics, and imagery layers. Repeat as necessary to describe maps and content at different scales or for use on different platforms, such as mobile devices.
- **Schematic requirements,** which include examples of the type of schematic output required; this goes on the map requirements form. If this is not applicable, note "NA."
- **List output requirements,** which include details of any information to be presented in the form of a report, list, map pop-up windows, or table, including headings and typical data entries.
- **Document retrieval requirements,** which consist of existing electronic documents that need to be incorporated into the information product, for example, textual information in the form of Adobe PDF, Microsoft Word, or text files; images and videos.

(continued on next page)

they display tabular data, the document retrieval form if they display images or other files. Or, one of the maps required for an information product could be a schematic, so your schematic requirements would be specified on the map requirements form. Essentially, you need to document everything you want in your information product, by means of a sketch, list, table, or sample of some sort—whatever effectively conveys the need.

Not every information product will contain all of these components, but they are listed here individually because considering whether each is needed or not is part of a sound planning methodology. Later on, IPDs will help you to specify what is required to make each information product and to verify that each contains all the needed elements once it's created. So you want to be sure your descriptions are thorough.

You can download IPD templates from the supplemental DVD.

FOCUS continued

- **Steps required to make the product,** which include details of the data elements and software functions needed to generate this single information product. Remember to include any steps that involve functions using mobile handheld devices.
- **Processing complexity,** which considers the number of complex geoprocessing functions required to create the information product each time it is generated.
- **Display complexity,** an assessment of the display rendering time and client traffic load, determined by number of features, layers, and sophistication of the output (for example, a detailed raster analysis output layer takes longer to display than a large-scale map of property boundaries).
- **Frequency of use,** which is an estimate of how often the product will be created and by how many people each year.
- **Logical linkages,** which include details of any linkages that need to be in place or be established between data elements in the database.
- **Error tolerance,** which is an estimation of acceptable levels of error in the information product.
- **Wait and response tolerances,** which include network timing issues and user demands; time requirements.
- **Current cost,**which describes the costs of producing the product using current methods.
- **Benefit analysis,** which describes the benefits to your organization of having this GIS-created information product.
- **Sign-offs,** meaning the signature of the person requesting the product on the last page (after initialing all the pages) to signify approval of the IPD, and the signature and initials on the benefits pages of the head of the department to which benefits from the information product will accrue. (Although not really a descriptive component, these signatures are so important to the IPD's usefulness that they are included here.)

Title

The title should be a precise, pithy, two- or three-word name for the information product, developed from the list of possible names collected at the technology seminar. The title must identify what the information is used for in a way that a layman can understand: in other words, name it "Sewer Backup Map" rather than "Engineering Situation Analysis."

Name of the department and person who needs it

Associating the department and requesting person's name with the information product encourages them to take responsibility for developing it into what they really need. Placing his or her name, along with the title, on every page of the IPD reinforces ownership of each information product. If a department manager requests a product, find out who will actually be using it. This is the name of the person that should go on the IPD. No end user name, no information product.

Synopsis

This is an easy-to-understand explanation of the information product and its purpose. This succinct narrative summary is usually the first page of the IPD. The synopsis should also briefly explain how the information product fits into the organization's strategic plan and supports the objectives of the management accountability framework.

Map output requirements

This section describes each map required in the output (if a map is indeed required), both visually and by listing its features (legend). It is important to include some sort of hand-drawn sketch or perhaps an actual example of the desired map previously prepared. The sketch can be simple but should show at least one of every feature type the final product is expected to display. If users need the map at two or more different scales, indicate this as well (draw both versions).

The sketched map or example should include the following information:

- Map title
- Vector layers that will be included on the map such as parcels, streets, buffers, or utility lines
- Raster layers that will be included on the map such as satellite imagery (orthophotographs) or continuous thematic data
- Legend showing how data is presented thematically on the map
- Any special symbology required
- Colors desired
- Scale bar
- North arrow
- 3D representations or schematics, such as utility networks
- Pop-up windows associated with map features, such as feature attributes, other associated data, photographs, videos, web links, sketches, and so on

Repeat as necessary to describe maps and content at different scales and for use on different platforms, such as mobile devices.

In addition to the visual features, it's important to describe any complex behavior you want the map to have, such as the ability to track temporal events, analyze a network, or maintain topology. You should also indicate whether map layers that will be served on the web are base layers (or "static layers"), thus able to be cached for faster performance, or if they are operational ("dynamic") layers that will need to be updated regularly. (Read more about data caching in the Focus box on page 74.)

Schematic requirements

An information product may present data in the form of schematics for ease of understanding. Schematics provide new ways to see and analyze relationships. These diagrammatic representations can approximate the real world in many ways: by using geographic coordinates, with a geoschematic emphasizing the topology of a network or other system, or as a pure schematic showing only the flow paths as linkages. A water system geoschematic connecting the fire hydrants, for example, could speed the response to a large fire by quickly showing firefighters where to draw water. Or, using schematics to trace the spread of disease with schematics has clear value in epidemiology. The latest software advances make creating and manipulating schematics easier, with many options for the user. Figure 6.1 shows three options for representing a sewer network.

List output requirements

An information product is not always a map. It could be simply a list of figures, a table, or a report; any of these could be stand-alone or part of an information product that also includes a map (e.g., a list of data associated with a feature is called an attribute table; list data can also be placed in a pop-up window so when a user clicks a feature the associated attributes appear). All these, along with spreadsheets, databases, and other text files, fall within the broad category called *tabular data.* Identify each of these lists, tables, or reports with a title, appropriate column headings, typical entries, and an indication of the source file for the data.

If predefined report formats exist, identify them as well. List data items can be moved into the desired report format. (Keep in mind that report creation capabilities in a GIS should be flexible enough to accommodate change in the future.)

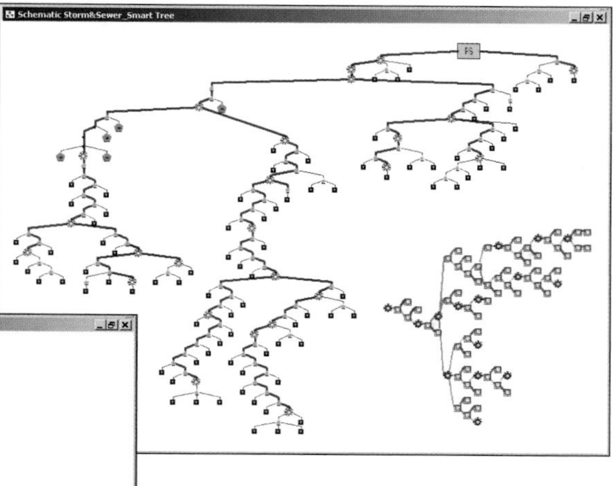

Figure 6.1 **These schematics show different diagrammatic representations of the same sewer network to help see and analyze connectivity.**

FOCUS

Three-dimensional representation

Significant steps have been made toward creating realistic three-dimensional (3D) representation and symbology. We are fast moving to a 3D GIS world. Already, databases can be extended to include geometry models for textured 3D solid objects representing buildings and other objects in the landscape. The ability to show 3D symbols of the basic shapes of points, lines, polygons (spheres, cubes, strips, dotted lines, and so on), and realistic out-of-the-box texture patterns exists. Extensive libraries of 3D objects to simulate 3D space are available, along with translators from common 3D object formats, such as OpenFlight, 3D Studio (3D models with a CAD orientation), and VRML (virtual reality modeling language). Symbology attached to these models can be geospecific in the shape of a 3D marker symbol or geotypical in the shape of a feature in a feature class. And 3D symbology is no longer merely speculative—using elevation measurements stored in a digital elevation model (DEM), you can convert 2D features into 3D features, allowing realistic display of building height, for example. In short, three-dimensional representation is becoming easier, quicker, and more flexible, and the foundation is being built for future development.

GIS can apply vertical exaggeration to satellite imagery (like the image of the Himalayas, top), achieving a 3D oblique imagery effect. Beyond the visual appeal, 3D data can be edited and analyzed with the right GIS software; for instance, streetlight luminosity is modeled using 3D visualization (center). With a unique city modeling and design software application (for example, CityEngine), you can create virtual cities like the one shown here (bottom).

Himalayas basemap imagery courtesy of NASA and I-Cubed (© 2009 i-cubed); streetlight and virtual city images courtesy of Pictometry International Corp., © 2012.

Document retrieval requirements

Text-based documents of a more narrative nature than lists often compose an important part of an information product, too; things like Adobe PDF documents, Microsoft Word documents, and plain old reliable text files. Documents included with information products may also be image files, audio or video files, scanned drawings, diagrams, or 3D representations. Text and image retrieval requirements may be specified on a single sheet, if feasible. In this descriptive category, include the likely number of pages per document to be retrieved from the GIS (typical and maximum), their typical file size, the key identifiers that will be used to search for and find the documents (an address, for example), and what you expect to see on the returned document (include a sample if possible). Where appropriate, indicate the need for document display. Specify whether users need to view the document, copy the whole (or portions of) the document, edit, or print the documents. If the documents are stored on the web or a server, include the URL (uniform resource locator) or directory path where they can be found. Documents may be stand-alone products or they may be part of another product, such as a map pop-up window. (The pop-up window can have links to related documents.)

You can add a map pop-up window or custom button to your desktop GIS application for quickly retrieving stand-alone documents.

Steps required to make the product

The components of the IPD described above (map output, schematic requirements, list output, document retrieval) clarify details of the information product required. Once you know this much about the information product, you can begin to evaluate the steps needed to make it. These steps should account for both the input data elements and the system functions required to make the product.

As you start to examine datasets, you'll notice that sometimes different names are used in various departments to refer to the same dataset. Carefully study the files and naming conventions and establish one standard name for each dataset. You may even have to create a data dictionary just to sort things out. It is good practice to settle on a single, simple, descriptive title that is unique to the dataset. You will use these titles later to create a master input data list, which will make use of much of the information you are gathering at this IPD stage of planning.

A *data dictionary* is a catalog or table containing information about the datasets stored in a database.

You should now write a step-by-step description of how to make the information product, in the order of the manual steps (many of which will be automated later) required for the GIS to provide the thing you want. This procedural description covers all stages, from the initial request to completion of the product. Of course, with GIS there are often several ways to do something, but for now all you need do is specify one logical and direct method. This will suffice to clarify the thinking about the user's real objectives and identify all the data required. Later on, the system will be fine-tuned to create an application that makes elegant use of the system functions.

Often, the most effective way to identify the sequence of steps needed to make an information product is to think of how you would create the product by hand using hard-copy data sources. For instance, it's logical to first list the steps for preparing the base layers before those of the operational layers. Setting down the steps for this method is not difficult, but it requires logic and a linear thought process. The key is to stay focused on the end product you envision and not to get distracted by other possibilities inherent in the data or by functions that you don't need. Rigorous care at this stage will often resolve questions about data availability and suitability, and identify sources of potential error.

Understand that the person who wants the product on his or her desk is the one who should create the description, including the steps to create the product. If they are not familiar with GIS functions, you may assist them, but do not do it for them. The desirable outcome of the planning process is for staff to be able to describe any future output they need from the GIS, and not be reliant on a GIS "priesthood."

For an example of how one might chart the steps required to make an information product, take a look at figure 6.12 on pages 63 and 64. Here are some additional tips that will help make the step-by-step description easier to write and more useful:

1. Use one dataset at a time and use its standard name. Write down the scales of any source maps that will be used as a reminder of the source data's resolution.
2. Use the generic function descriptions provided in the lexicon at the back of this book. They are easy to understand and will translate into any system-specific piece of software.

The GIS manager works closely with the person who needs the product, to best determine how it should be created, but the steps must be spelled out by the user.

3. For every time a function is used, explain the work it's doing on the dataset and clearly identify the specific data elements (layers, features, attributes, relationships, documents, images, etc.) accessed in the process. This is very important, as the dataset could contain many more data elements than the ones you need, and you must be sure to work with the right one(s). You will also cross-check the product output requirements against the series of steps on your chart to make sure that all the data elements needed in the output product have been produced by the functions working on the datasets.
4. Always operate on the assumption that the results of using any function are available for use in later steps. For more complicated processes, you can represent the steps visually in the form of a flowchart to aid in communication. The steps to make the product should be clear and should help identify the data needed, functions invoked at each step, details of any interim products, and the final product. Complex processes will likely be executed using a geoprocessing model, and you can't create the model without methodically spelling out each element and function.
5. If data is not readily available in-house, now is a good time to think about how you will obtain it—from the government, a data vendor, a data clearinghouse website, ArcGIS Online (see Focus box), or in-house creation.
6. If the organization's staff is not aware of the range of GIS functions available or the terminology to describe them, they may need to seek further training in basic GIS functionality.

Don't overlook the capabilities of mobile devices in the array of functions at your fingertips. Many handheld devices and Tablet PCs with differing capacities and operating systems are able to interact with GIS. You need to consider these mobile devices both in terms of communications limitations as well as capabilities. And this stage of the process is the time to do it so that you'll know, for example, if an application needs to be available on desktop PCs as well as handheld computers or smartphones. These devices offer important new capabilities for GIS developers, including the ability to operate through loosely coupled architecture for disconnected editing of the central database in the field. You can read more about mobile GIS technology in chapter 7.

Processing complexity

You'll need to assess the complexity of the geoprocessing functions used to prepare, analyze, or display the information product. Processing complexity is either high (four or more complex functions) or low (zero to three complex functions). Complex functions are noted in the lexicon (in the back of the book) by an asterisk. (For example, any functions performed using ArcGIS Spatial Analyst

FOCUS

ArcGIS Online

Your organization may want to take advantage of the wide array of cloud-based basemaps and imagery offered through ArcGIS Online. Doing so will significantly reduce data storage and processing loads, as well as hardware and administration costs. It also improves display performance because the basemaps are automatically cached for the user.

From ArcGIS Online you can choose from aerial satellite image maps, street maps, political boundary maps, topographic maps, landscape series maps, or any number of community-sourced basemaps, all of which are available for standard usage. ArcGIS Online also offers worldwide geocoding and routing services, maps, and apps when a subscription is purchased. Find out more at `http://www.arcgis.com`.

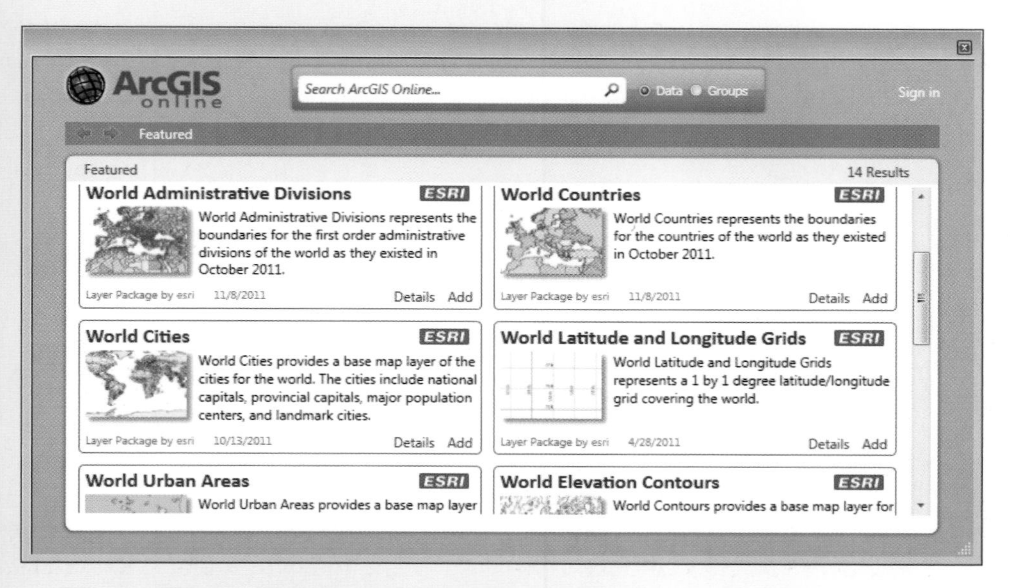

You can choose cloud-based ArcGIS Online basemaps for your web map service information products. Esri's ArcGIS Online

FOCUS

A rapid prototyping tool

When datasets are already entered into the GIS, recent advancements can make the development and execution of prototype GIS applications far quicker and easier than ever before. Wizard-based and graphic-based construction of GIS applications is possible using a wide variety of data types and geoprocessing functions. A rapid prototyping tool, such as Esri's ModelBuilder tool (embedded in all license levels of ArcGIS for Desktop software), gives you the advantage of combining datasets already entered into the system with a series of geoprocessing steps to produce information products. These products can be simple or complex; in fact, they can be extremely complex. They can be combined models. The output can be examined as the model goes through several stages of development. With a tool like ModelBuilder you can use and combine a broad set of geoprocessing tools to compile and manage data, perform complicated analyses and iterative processes, extract data, perform cartographic functions, and automate frequently performed workflows. Models can then be shared and published to facilitate collaboration within and between organizations.

Rapid prototyping tools will continue to play an important role in iterative design because a model can be shared between people who can add to it, revise it, and implement it. An accepted model can be published, rather in the way an application is currently published, but with the steps in the model graphically illustrated and amenable to revision.

Future directions of development may lead to building interfaces between geoprocessing models and business workflow applications (such as Visio Enterprise, iGrafx FlowCharter, or Workflow Analyzer), wherein the models can both construct the information products required by the business workflow and interface with CASE tools in physically designing the databases the models need.

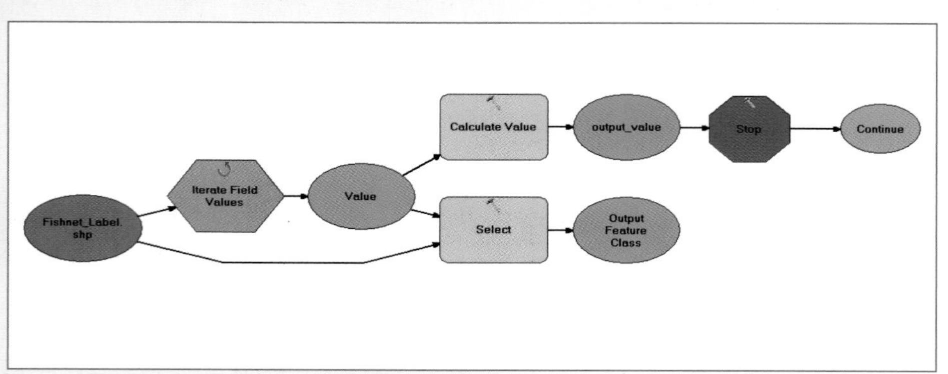

A geoprocessing model automates multiple GIS functions.

would be considered complex.) Include batch processes that are performed in the background (caching base layers, for example), as well as those performed in an active client session, such as generating a line-of-sight layer on the fly or running a complex process model.

Processing complexity has a significant impact on software selection and platform sizing, to be discussed further in chapter 9. Now is the time to assess the complexity of your needed information products, not later.

Display complexity

You also need to assess the display complexity if the product is a map or imagery, especially if it will be served on the web. Display complexity estimates the processing load required to generate the user display during an active client session.

There are two parts to display complexity. The first part is an estimate of the effort required by the computer to generate a single display. The second part is an estimate of the display traffic (amount of data delivered over the network to the end-user display).

For display processing complexity, assign a degree of *light, medium,* or *heavy.* This is based on the following criteria:

- Multiple service mashups (light 1–2, medium 3–5, heavy 6+)
- Number of layers (light: less than 20; medium: 20–40; heavy: 40+)
- Number of features per layer (light: less than 1,000; medium: 1,000–10,000; heavy: 10,000+)
- Display quality enhancements such as anti-aliasing, transparent layers, dynamic Maplex labeling, special effects (fill, backgrounds, callouts), very large text (60+ point), or geoprocessing outputs such as hillshades or networks (light: none; medium: 1; heavy: 2+)

For display traffic complexity you have two possible deployment patterns to consider: GIS desktop applications hosted on Windows Terminal Servers (WTS), and web services hosted on a web mapping server.

- WTS traffic requirements are either low or high based on the type of display (light for vector only, heavy if the display includes raster imagery).
- Web services traffic requirements are either light, medium, or heavy based on the following considerations:
  - Number of image services (overlays) supporting a single display (light: 1; medium: 2; heavy: 3+)
  - Output image format (light: PNG8 vector only; medium: JPEG/PNG32; heavy: PDF)
  - Size of client display resolution (image pixels)—light: less than 600 × 400; medium: 1024 × 768–1200 × 800; heavy: 1280 × 1024 or higher

It's important to evaluate the display complexity of your information products early in the game because it affects platform selection and sizing, especially for distributed web services (discussed in chapter 9 and appendix D). Use the highest level for any of the criteria above as your assessment rating.

Frequency of use

You also need to estimate the overall demand for the information product—how many times per year will it be generated? The frequency may range from once per year (a map to hang on a wall) to ten thousand times per year (some emergency routing products). Write down the numbers of maps you anticipate producing, lists you think you will generate, and documents you will retrieve each year. Then, for the IPD, try to project these numbers out to five years. Include a simple table in the IPD summarizing these frequency-of-use estimates.

After determining all of the functions and when they are used in the process, you can count the number of times each function will be used in the creation of the information product. You may have some products that require multiple functions, and those functions may

FOCUS

Measuring display complexity

Tools are available for measuring display complexity once you start building and deploying your information product. Display processing complexity is a measure of display rendering time. Traffic complexity is a measure of client display traffic (how much traffic is sent to the client for each display). GIS desktop map publishing tools can be used to measure map document rendering times. Standard system performance monitoring tools, such as Windows performance monitor and Fiddler, can also be used to measure rendering times as well as client display traffic. Rendering time will depend on the complexity of features and the speed of your computer; older computers will take longer to process your display than new computers. More details on display complexity and how it can be measured are provided in appendix D.

have to be invoked many times. After you determine the number of times per year the product is generated, you can multiply that by the number of times a specific function is used to make the product. The result is the number of times the function is used in a year to make that product (figure 6.2).

You are now getting a first view of how often you will use the functions in your system. On completion of your first round of IPDs, you can create a list of the functions that will be required from the GIS overall. Such a list will be useful in evaluating your choices among potential GIS software packages and in scoping your system requirements for hardware, as you'll see in chapter 9. The summary of totals, along with the display and processing complexity evaluation totals for all IPDs, become important tools for ensuring that you plan for a system that will effectively and efficiently meet your needs.

Logical linkages

The next step in describing an information product is to determine the relationships needed between data elements and between datasets. These relationships are called *logical linkages,* and they must be in place when you build your database later on. In the IPD, you need to establish how data from different datasets will be able to be connected to create the end product.

There are four types of logical linkages:

1. Relationships between the graphic entities of features (points, lines, polygons) and lists of their attributes (e.g., name, code, description, address).
2. Relationships between maps or map layers: these are relationships between the different kinds of maps (or data layers) that you will use. (For example, can they be overlaid? Are they the same scale and map projection?)

Number of times a product is created in one year	X	Number of times a function is used during creation of the product	=	Number of times a function is used to create the product in one year

Figure 6.2 **Equation to determine frequency of function use.**

3. Relationships between graphic entities of features. (For example, line-to-line linkages to create the topology of a network suitable for network tracing and analysis.)
4. Relationships between attributes: these are relationships between characteristics of data, stored in either spatial or nonspatial attribute tables. (For example, when identifying a property boundary, what other information about that boundary needs to be displayed?)

Don't look for every linkage you can establish, but do look for linkages you need that might not be there. When you perceive necessary logical linkages not currently in the source data, you're identifying a problem that must be solved before this information product can be made. If, for example, you know that you need a linkage between a house and a sewer, it needs to be in the database. If you find out after you've built your database that you need such a linkage between each house and a sewer, you might have to revisit every house and sewer to input the data. This may not be significant with a town of only 500 dwellings, but if you have 120,000 dwellings to revisit, the problem becomes onerous indeed. Information about linkages becomes more significant as the size of your database increases.

Error tolerance

Error tolerance is the amount of error that you can accept in the information product. The question you must answer is, "How wrong can it be?" When someone requests an information product, it's because they expect it to provide a benefit, such as saving staff time or increasing operational efficiency. If the information product contains errors that require time to correct or another method for determining the right answer, you receive neither the savings in staff time nor improvements in efficiency.

The IPD should address possible errors, the result of errors, the impact on benefits, and the amount of error that is acceptable. That is, you must establish how much error can be tolerated in the information product without losing its usefulness.

It is important for users to think about error and consider what level of accuracy is important in terms of cost versus reliability. The user's view of their own needs is their bargaining position in the discussions on data quality during implementation. Once costs and benefits are weighed, users may not get their desired accuracy, but at least they'll be prepared for this and cognizant of the degree of reliability they can count on in the information products.

FOCUS

Maximize efficiency

To maximize efficiency, you want to minimize processing load. To do so, you must precompute as much as possible. This is an important consideration as you outline the steps for creating information products. For instance, if there is a base layer that will be used in several web map services, you should cache the layer for faster display (see "Focus: Using and creating a data cache" on page 74). If creating an information product requires several geoprocessing functions, plan to preprocess as much as you can; then you may only have to enter a variable and run the remainder of the model. This will save time and computer resources. For map creation, use templates as much as possible so you don't have to start from scratch every time. You can also create data model templates, so if you receive or create new data there you can store it in a preexisting organizational structure. In short: maximize efficiency whenever and however you can.

The levels of quality control established to ensure that data meets user needs can significantly increase costs. Achieving the elusive final 10 percent of accuracy may cost 90 percent of the total costs of building the database. With users, you should determine the maximum amount of error that can be tolerated while still sustaining the intended benefits from the information product. You need to find out where the balance is between accuracy requirements and data-entry and quality-assurance costs.

There are four types of error:

- Referential: an error in a reference to something, such as a wrong address, label, number, or name.
- Topological: a linkage error in spatial data, such as unclosed polygons or breaks in networks.
- Relative: an error in the positioning of two objects relative to each other.
- Absolute: an error in something's true position in the world.

Error tolerance is expressed numerically but differently depending on the type of error, as you can see in figure 6.3. Usually, referential and topological error tolerances are expressed as a percentage of total occurrences, while relative and absolute errors are expressed in linear distance.

Figure 6.4 shows examples of each type of error from various information products.

In thinking about data, remember there is no such thing as inherently "good" data. In working terms, there is useful data and useless data. Data carries no intrinsic value or accuracy. There is no accuracy that data should contain in theory—only the accuracy that affects people doing real work; in other words, the accuracy demanded from the data to create a useful information product. Data remains useless if its accuracy is not related to the reality of what the data is being used for.

You can manage the data quality control process with a data review software product that facilitates error-checking and correction. You can minimize topological errors by establishing topology rules for your GIS data.

Read about error as it relates to map scale and resolution in chapter 7.

Test your understanding of possible data errors: take the Error quiz on the supplemental DVD in the back of the book.

Wait and response tolerances

Wait and response tolerances are related but distinct concepts that must be included in the IPD. Wait tolerance is a measure of how robust the computer and network system must be, that is, the maximum allowable time (with the computer up and running) between the last keystroke and the full display or output of the information product (figure 6.5). For some emergency dispatch uses, the wait tolerance can be as low as 0 to 1 second (specified

Error type	Specific error	Tolerance
Referential	Incorrect street address	2%
Topological	Incorrect break in street network	0%
Relative	Sewer line shown on the wrong side of the street	± 2m
Absolute	Floodplain not aligned with property boundaries	± 10m

Figure 6.3 **Expressing error tolerance.**

in subseconds); for other applications it can be up to an hour, or so high it's almost irrelevant.

Wait tolerances can influence how you design your computer network, while response tolerances have more to do with the human process. Response tolerance is the maximum allowable time between the arrival of the request or critical data at the GIS office (or wherever processing takes place) and the delivery of the information product to the user. Gauging how much time can be spent in responding to the request for an information product will help you understand how many GIS office hours, staff members, and so on, will be needed to meet

Referential error	
Possible occurrence	Incorrect street address.
Results of error	Delivery to wrong property.
Impact on benefits	Time wasted during delivery reduces benefits of having product. Pizza may get cold.
Concerns for error tolerance	Must balance data entry and checking costs with amount of acceptable error. 2% of addresses wrong.

Topological error	
Possible occurrence	Incorrect street network.
Results of error	Route may take longer than necessary.
Impact on benefits	Loss of life or property because an emergency services vehicle is delayed.
Concerns for error tolerance	If loss of life may occur, zero tolerance may be required.

Relative error	
Possible occurrence	Sewer line shown on the wrong side of the street.
Results of error	Excavation for sewer repairs in the wrong place.
Impact on benefits	Increase in on-site costs.
Concerns for error tolerance	How much error can the sewer line have before it's ineffective? It might still be effective if it is only a little off-center, but if it is on the wrong side of the road, it's ineffective. ±6' (2m).

Absolute error	
Possible occurrence	Floodplain boundary not aligned with property boundaries.
Results of error	Uncertainty of location. Are you in the floodplain?
Impact on benefits	Owners paying for flood insurance when it is not needed. Owners not being protected when they are in a floodplain.
Concerns for error tolerance	Need to balance costs of data, costs of checking, and the amount of acceptable error. ±30' (10m).

Figure 6.4 **Examples of the four types of error in information products.**

Category	Range	Units to specify
X-Low	0–0.5 seconds	Sub-seconds
V-Low	0.5–10 seconds	Sub-seconds; seconds
Low	10–60 seconds	Seconds
Medium	1–5 minutes	Minutes
High	5–60 minutes	Minutes
V-High	1 hour +	Hours
Non-critical	N/A	N/A

Figure 6.5 **Wait tolerance categories for full display of information product, measured from last query keystroke.**

the demand. You know the response tolerance is set too high, for example, when the forest is burning on Sunday but the maps aren't available until Monday!

Current cost

The next step in building an IPD is to document what it currently costs you to create the information product without GIS. Your estimate should include both labor and materials and account for how many times a year your organization incurs this total expense.

The cost figures you calculate can be used in cost-cost comparisons to help justify the implementation of the GIS. For example, if an information product costs a few hundred dollars and is used only once or twice a year, the automation costs will not be recovered in a reasonable amount of time. On the other hand, if you have an information product that takes a lot of staff time to create and is needed frequently, the current cost will probably justify automation. This step helps you identify the level of effort required for each information product being requested. It also helps explain to management the magnitude of the work being attempted by the GIS, and sometimes explains why the information product is not made using current methods. If the information product is a brand new undertaking that is not currently being produced manually, you can estimate how long it would take without GIS, or you might just skip this step.

Benefit analysis

For the final step in preparing the IPD, the person who needs the information product on his or her desk should perform a benefit analysis. Also, the GIS manager should have a grasp on what benefit will result from the information created through GIS. You should be able to weigh the costs of system and data acquisition (the input) against the benefits your organization expects from the output. Consider the following three categories of benefits:

- Financial savings: actual cash saved from current budgets if the GIS made the required information products (i.e., reductions in current staff time, increases in revenue).
- Direct benefits to the organization: beneficial results of implementing the GIS, which were not available before. These could include improvements in operational efficiency and workflow, or the reduction of a liability, for example.
- External benefits: benefits that accrue to others who are not directly using the GIS. For instance, the general public benefits from lower fire insurance rates, indirectly, when better fire response times result from the fire department's immediate access to reliable maps.

Once you've added up the benefits of each information product, you can compare the benefit total with the cost of implementing the GIS, including acquiring data needed for the information products and the staffing required (see "Focus: The cost model," page 162).

Calculating benefits always takes thinking, but is critical in the GIS planning process. At one time, most economists held the view that the only legitimate value that could be attached to data was the value someone was prepared to pay for it. In practice, this is ludicrous because no one buys data just for the sake of buying it. People buy data because they want to do something with it that is going to create information that will benefit them. We need new models—methods to calculate the benefits that accrue from having the information.

You will find a methodology for measuring the benefits of information—benefit-cost analysis—described in chapter 10. The methodology has been tested and proven successful in Canada, Australia, and in state and federal governments in the United States. The organizations that used the methodology accepted the outcome and the results from it, which is the real measure of success.

Take the Benefits quiz on the supplemental DVD to practice matching specific benefits with the correct benefit category (saving, benefit to the agency, or future and external benefit).

Sign-offs

After each IPD is complete, ask the person writing the description to review and initial every page—and to sign the last page—of the IPD. This sign-off on each page confirms that it accurately describes the information product requested. If there is no signature, there can be no information product. It is best to make this clear from the beginning, so the person responsible can make an effort to ensure that the final description meets the need. People sometimes don't realize the amount of work that goes into building a GIS, so it's important to make clear to them early on that their organization is going to be spending resources on their behalf, and it is imperative to spend it on information they truly need.

Another person's initials and signature are essential as well: the department head overseeing the budget to which the benefits will accrue must initial the benefits pages and sign the last page. If he or she will not initial the benefits pages, find out why and make the needed revisions. If you can't reasonably accommodate the changes, then there is a fundamental problem with the information product.

Again, no department head signature, no information product. This harkens back to the first recommendations of this methodology—be sure to gain upper management's full support for the planning from the beginning of the project, and regularly inform management of progress. A consistent and rigorous approach to approvals is what makes GIS successful in established organizations. And the time-honored way to win approval is to demonstrate how benefits will outweigh costs quickly and time after time.

Before moving on to the case study, here are some final thoughts about creating IPDs. Much of the information gathering and writing of an IPD is best done by the person requesting the information product, with the help of the GIS planner or planning team. Creating the IPDs is probably the single most important task in the entire GIS planning process, and the most fruitful. Not an ounce of effort is ever wasted in thinking about the nature and details of desired information products. At some stage in running a GIS, all of the information in the IPDs will be needed by the GIS manager to plan his or her day-to-day work. Sooner or later, the thinking is going to have to be done to explain the work to users and senior management and to run the system itself. Doing this thinking when faced with daily production demands in a production GIS shop is a recipe for GIS manager madness (or at least frustration and overwork). It is better to do it during planning, before production begins and not after.

Case study: Tracking the IPD

This section tracks the creation of an information product description, step by step.

Gary Jones works in the engineering department for the City of Jackson. Part of his job is to oversee the maintenance and repair of the sewer system. He needs an information product that will help his staff deal with

Title: Sewer backup system
Required by: Engineering department
Name: Gary

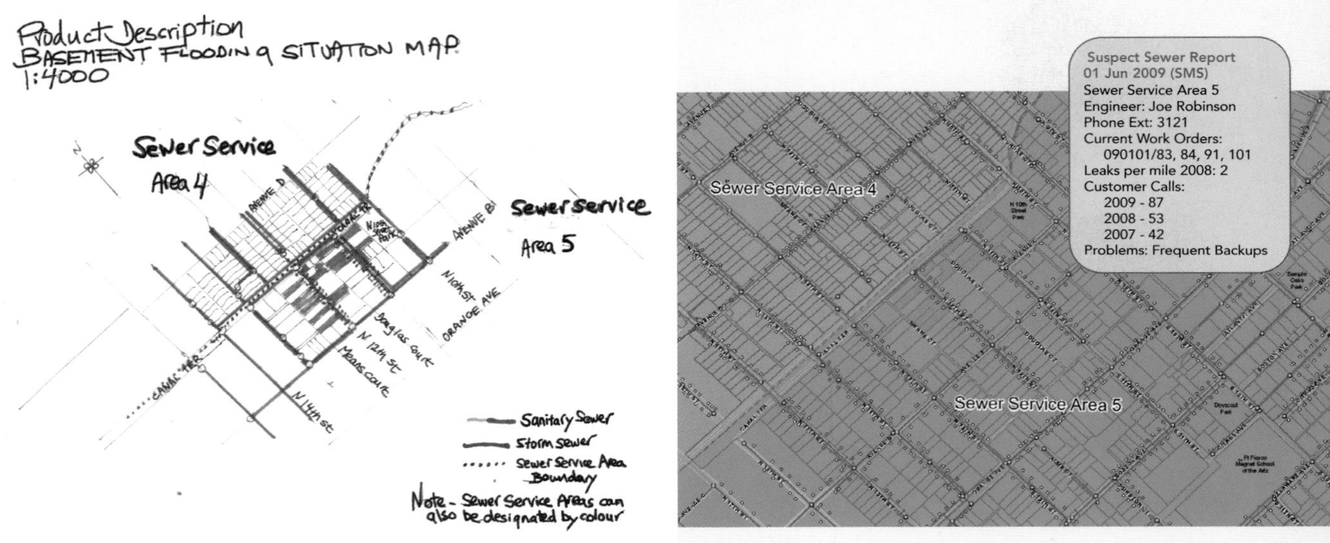

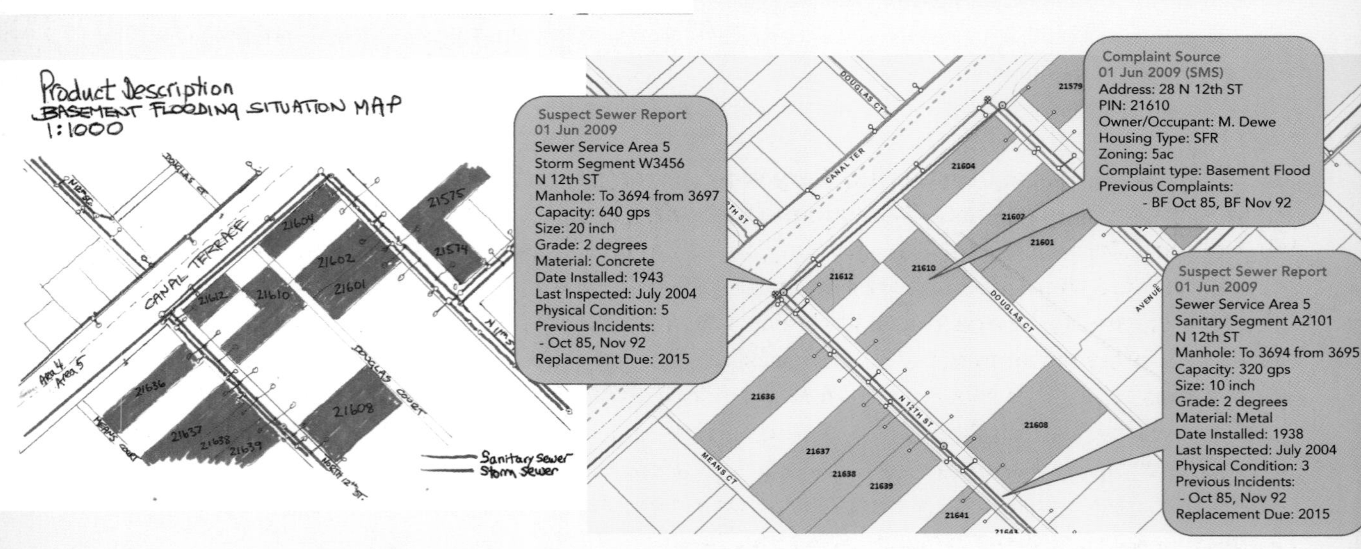

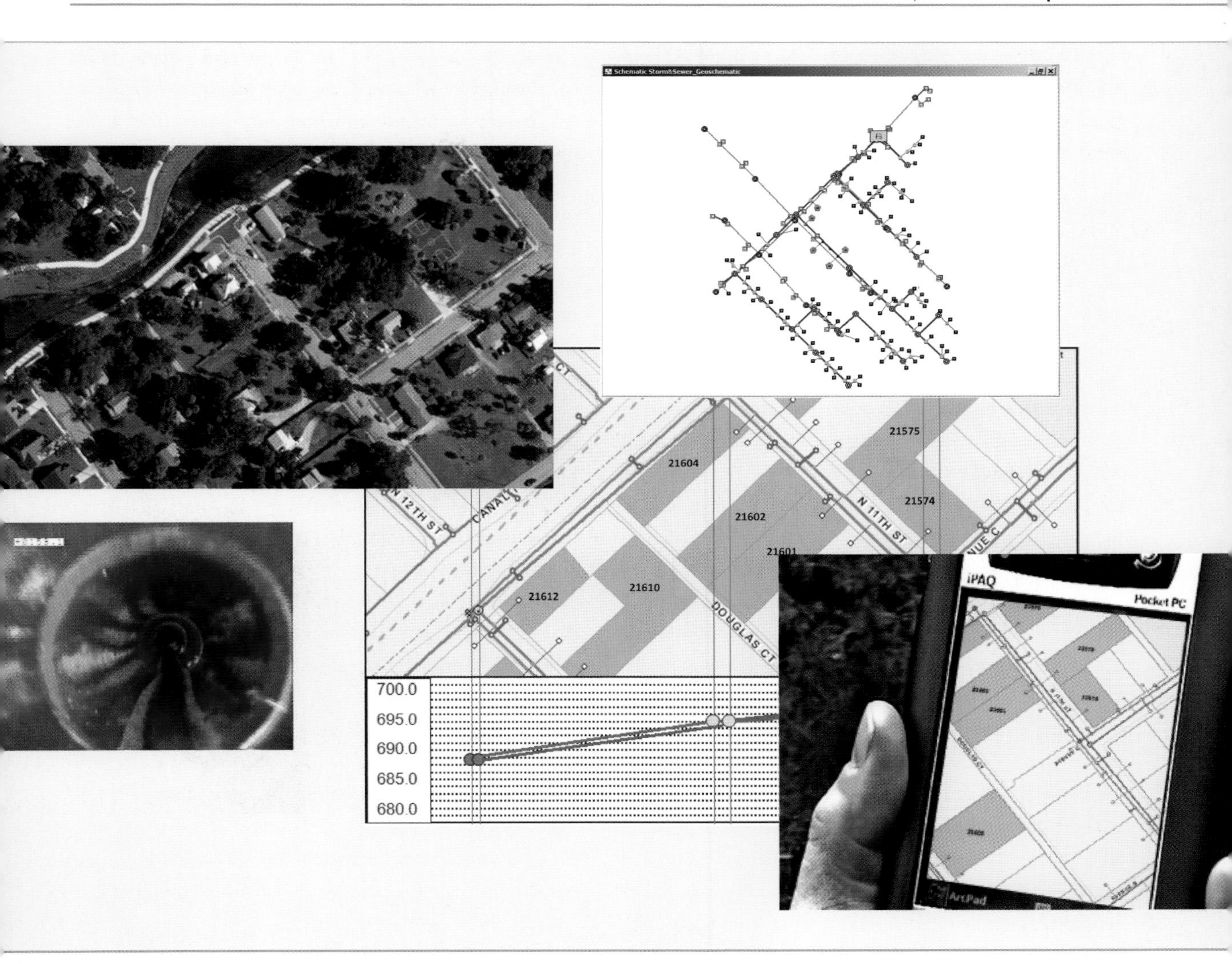

Figure 6.6 **The information product Gary wants is multifaceted. It includes a map with details at two scales. (Shown here are the hand-drawn maps that Gary sketched, as well as the finished GIS maps. Your IPD may only have sketches at this stage.) The maps must contain vector data for streets, parcels, storm and sewer lines, and manholes; pop-up windows that produce on-the-fly reports showing relevant attribute data; and a satellite image basemap layer. The information product must also be able to generate schematic diagrams of the sewer lines, as well as retrieve the following documents: image layers (such as a street-view photograph, not shown here), the related engineering plan and profile (a graph showing the elevation of the sewer lines); and the video file from the latest inspection of the suspected sewer line. Finally, the information product must be capable of receiving updates from inspectors using mobile devices in the field.**

frequent sewer backups and subsequent basement flooding. During his city's recent technology seminar, Gary requests a basement flooding situation map as part of the information product he needs the GIS to produce.

The background is as follows: In this city the sewers are generally known to be in a terrible state. The original lines date from the middle of the nineteenth century. At one time they were combined sanitary and storm sewers, but were separated later on. Now the city's engineering department is dealing with frequent complaints. The aging and modified sewers tend to flood, filling people's basements with sewage—a very unpleasant situation to say the least. Every year there are fifty sewer backup incidents, on average, all over the city. This is not a trivial situation. When a backup occurs, homeowners telephone city hall for action, and the engineering department must react quickly.

Gary needs to be able to access the information product that displays the information he needs about a particular area. But before the GIS can create it and make it available, Gary will have to describe it thoroughly in terms of the components required for it to be built into the database. He will have to write an IPD, engaging your help as a planner.

The first components of this information product description will appear on every IPD form, as shown in figure 6.6: title or name of the information product, the department requiring it, and the person who needs it. As the person identified on the IPD, Gary is responsible for making sure the IPD accurately describes the information product he needs. Once the IPD is complete, he will need to initial each page and sign the document, verifying that the description is correct. The head of the engineering department will have to sign the benefits pages of the IPD as further verification that the information product as described is exactly what they need.

Now we'll examine in detail the other components of Gary's IPD, beginning with a brief synopsis of the information product.

Synopsis

In a nutshell, Gary needs to be able to enter the address of the sewer complaint and have the GIS return an information product. He wants a robust map that provides information in several different ways. The map should provide a small-scale view of the city's sewer service areas, a large-scale view of the property or properties involved in the backup (showing parcels and streets), the storm and sewer segments under suspicion, the segments connected to them (both upstream and downstream), manhole locations, and any other potentially relevant landmarks in the immediate surrounding area. The map should have pop-up windows containing information about sewer areas, the complaint source, and the storm and sewer segments. The map should also contain an aerial satellite image of the affected area, as well as a photographic view (street-view) if available, so city workers can quickly find the problem area. Including labels identifying the lines and manholes that need to be serviced will help avoid common human error. An engineering plan and profile drawing (indicating elevation and slope of the lines), as well as schematic diagrams showing connectivity of the sewer lines, will be helpful for analysts and repair crews. In addition, Gary needs to understand the known condition of the suspected segment, so he wants the GIS to retrieve the latest inspection's interior video scan (video scans are standard inspection practice in Jackson). Finally, Gary wants the basement flooding situation map to be updated with new information as it's discovered, so inspectors will need to use mobile GIS to record new data and update the database in real time. All of these things are multiple parts of a single information product.

Gary's synopsis should briefly describe the information product (no more than a paragraph), and may also include a rudimentary mock-up like that shown in figure 6.6. Most important, the synopsis should address—in specific terms—how the proposed information product fits into the organization's business model and facilitates its mission. For example, you could put this part of the synopsis in boldface or in a shaded box, to bring it to the CEO's attention:

> This information product is central to **management accountability framework objective #12**—improve average problem resolution time of basement flooding by 20 percent in the next three years. GIS use will decrease the staff time and effort required for a satisfactory response to flooding situations by 80 percent, decrease the number and expense of litigation related to domestic and commercial flooding, and minimize environmental impact due to storm water release into rivers and streams. Budgetary impact will be $160,000 benefit per year to the sewer maintenance operation budget.

Map requirements

To fulfill the descriptive elements of this IPD component, Gary makes a sketch of the type of map he needs. Gary thinks that a basement flooding situation map should illustrate the following:

- Sewer service area: a small-scale map (1:4,000) identifying which area of the city's extensive utility network is affected
- A larger-scale map (1:1,000) showing the following:
  - Property numbers (addresses)
  - Locations of the houses whose sewers are backed up
  - Property ID numbers of the affected houses
  - Sewer segments the houses are connected to
  - Street names
  - Manhole numbers
- A satellite basemap layer of the sewer areas, in three scales (1:8,000, 1: 4,000, and 1:1,000)

Schematic requirements

Understanding that schematic outputs provide a different way to visualize and analyze relationships, Gary is also requesting that the GIS outputs schematic diagrams of the sewer lines, displaying connectivity and flow direction.

List requirements

Gary needs four lists. He makes note of all the information needed to build them. First, he needs a list of pertinent details about the operation of the affected area (figure 6.7):

- Service area number

Title: Sewer backup information product
Required by: Engineering department
Name: Gary

List #1							
	Sewer operation information						
Headings	**Sewer service area**	**Engineer**	**Phone Ext**	**Current work orders**	**Leaks per mile 2012**	**Customer calls**	**Problems**
Typical entries	5	Joe Robinson	3121	090101/83, 84, 91, 101	2	2012 - 87 2011 - 53 2010 - 42	Frequent backups
	4	Tom Rolofson	6120	090101/65, 72, 81	1	2012 - 37 2011 - 46 2010 - 48	None identified
Source	SIMS	SIMS	SIMS	SIMS	SIMS	SIMS	SIMS

Figure 6.7 **IPD format for List #1.**

- Engineer in charge of the area
- Phone extension
- Current work orders
- Leaks per mile (according to previous year's data)
- Customer calls (number of calls over the last three years)
- Known problems

The IPD entry format for this list is the same for the other three lists, all containing categories for headings, typical entries, and source information. In this case, since the list contains administrative details, the source for all items in the list is *SIMS* (sewer information management system), the sewer data file maintained in Gary's department on a PC.

The second list (figure 6.8) contains complaint source information. The data for this list comes from either the complaint file itself (a 5-by-7-inch card file); SIMS; or a city mainframe database of property development information stored by the assessor's office, called the property development information system, or *PDIS*. The list includes the following:

- Address of the affected property

Title: Sewer backup information product
Required by: Engineering department
Name: Gary

List #2									
	Complaint source information								
								Connection	
Headings	Address	Prop ID	Owner occupant	Housing type	Zoning	Type	Previous complaint	Sanitary	Storm
Typical entries	28 N 12th St.	21610	M. Dewe	SFR	5ac	Basement flood	Oct. 1992 Nov. 2004		
	26 Cooper St.	32968	M.A. Brown	SFR	5ac	Basement flood	Oct. 1995 Nov. 2004	To 07062	To 07062
Source	Complaint file	PDIS	PDIS	PDIS	PDIS	Complaint file	Complaint file	SIMS	SIMS

Figure 6.8
IPD format for List #2.

Title: Sewer backup information product
Required by: Engineering department
Name: Gary

List #3								
	Suspect sewer report							
				Manhole				
Headings	Storm or sanitary	Sewer service area	Segment #	To	From	Capacity	Size
Typical entries	Sanitary	5	N 12th St.	3694	3695	320 gps	10"
	Storm	5	W3456 N 12th St.	3694	3695	640 gps	20"
Source	SIMS	SIMS	SIMS	SIMS	SIMS	SIMS	SIMS

Figure 6.9 **IDP format for List #3.**

- Property ID number
- Name of the occupant/owner
- Housing type: e.g., single-family residential
- Zoning
- Type of complaint
- Date of previous complaints
- Details about the relationship between previous complaints and the sanitary sewer or the storm sewer (in other words, the suspected lines)

Gary needs a third list of details about the suspect sanitary or storm segment itself, including the following information (figure 6.9):

- Segment type
- Segment number
- Manhole numbers
- Sewer capacity
- Sewer size
- Sewer grade
- Construction materials
- Date of installation
- Last inspection date
- Class (physical condition)
- Date last cleared
- Number of floods
- Previous incidents
- Replacement due

The fourth thing Gary needs in list format is the items required for mobile data collection (figure 6.10). This list isn't displayed on the map; rather, it serves as a guide for field workers who are inspecting the basement flooding complaints. They will enter the data on a mobile GIS unit, and it will then be stored in SIMS.

Title: Sewer backup information product
Required by: Engineering department
Name: Gary

List #4					
	Data for mobile entry				
Headings	**Water depth during flooding**		**Rainfall**	**Source of entry**	**Basement elevation (height above sea level)**
	Manhole	**Basement**			
Typical entries	1.5 ft.	1.0 in.	2"/hr	Floor trap	320.0
				Walls	315.0
				Cleanout	350.0
				Depressed driveway	298.1
	2.0 ft.	4.5 in.	1"/hr	Depressed driveway	289.0
Source	Inspector		Inspector		SIMS

Figure 6.10
IPD format for List #4.

Grade	Material	Date installed	Date last inspected	Class (phys. cond.)	Previous incident	Replacement due
2"	Metal	1938	July 2009	3	Oct. 1992 Nov. 2004	2015
2"	Concrete	1943	July 2009	5	Oct. 1995 Nov. 2004	2015
SIMS	SIMS	SIMS	SIMS	SIMS	SIMS	SIMS

Figure 6.9 (continued)

Document retrieval

After describing the maps and lists, Gary specifies any documents he may need to retrieve. The engineering department employs a robotic video system capable of delivering video recordings showing pipe condition in the segment. These videos are all date-stamped and referenced to the segment number. (The form shown in figure 6.11 specifies, next to "Spatial," that Gary will retrieve these video files based on sewer segment number.) Gary tells you he needs to be able to observe the videos and transfer some still shots to hard copy without changing the documents on which they appear.

Gary also wants to be able to retrieve the engineering plan and profile of the sewer line in question, so he will submit a separate document retrieval form for that.

Inspectors may periodically record video scans of sewer and storm lines, to be stored for quick retrieval when there is a maintenance problem.

Title: Sewer backup information product
Required by: Engineering department
Name: Gary

Document retrieval			
#	**Dataset name:** Sewer characteristics (SIMS) file		
Document title: Sewer video scan file			
Number of pages per retrieved document		Typical 2	Maximum 5
Search keys (all) **Spatial:** Sewer segment number **Attribute:** ---			
Data elements (required to be seen) Sewer interior video scan (manhole to manhole) of suspect sewer segments			
Action: (Check as appropriate)	✓	Visually observe	Read only
		Copy whole	Hard copy
		Copy whole	Digital
Change: (Check as appropriate)	✓	Copy part	Hard copy
		Copy part	Digital
		Add data	Which elements
		Delete data	Which elements
		Edit data	Change errors
No change permitted	✓		

Figure 6.11 **IPD document retrieval form for the video scans, engineering plan and profile, photographs, building footprints, or any other document that needs to be retrieved. You can add a pop-up window or custom button to your GIS application for quickly retrieving documents.**

Finally, he wants to be able to easily retrieve existing street-view photography. (Only the document retrieval form for the video scan is shown here.)

Now that you know what Gary wants—maps, schematic diagrams, lists, images, videos, and plan and profile graphs—start thinking about how this information product, with its disparate data sources, can be produced:

- What data do you need to make the information product?
- Where will the data come from?
- Which functions are needed to create the products?
- How do the component parts relate to one another?
- What common data elements in the different datasets are needed to link them together?

This is where Gary needs your help and expertise as a GIS planner. He can tell you that the engineering department has sewer characteristics stored in the SIMS database on a PC, as well as paper maps of the service areas. City property data (PDIS) and satellite imagery need to be obtained, and sewer line maps need to be digitized. You can tell him that the sewer characteristics data should be linked to the property file to allow sewer segment numbers to be linked to street addresses.

Steps to make the product (data and functions required)

Together, you've figured out what you need to know about the information product envisioned in order to list the steps required to make it. Figure 6.12 gives details of the step-by-step process of building Gary's information product. From his description of the workflow, you identify the datasets (left column) and system functions (right column) required to make the product intended to facilitate the workflow. This organized document—the "steps to make the information product" table—is a vital component of the IPD.

The table's right column contains very important information. It specifies not only the system functions needed to create the information product but also every data element that must be taken from the dataset and included in the product. You'll use this later to confirm that all the required data elements will actually be in the product.

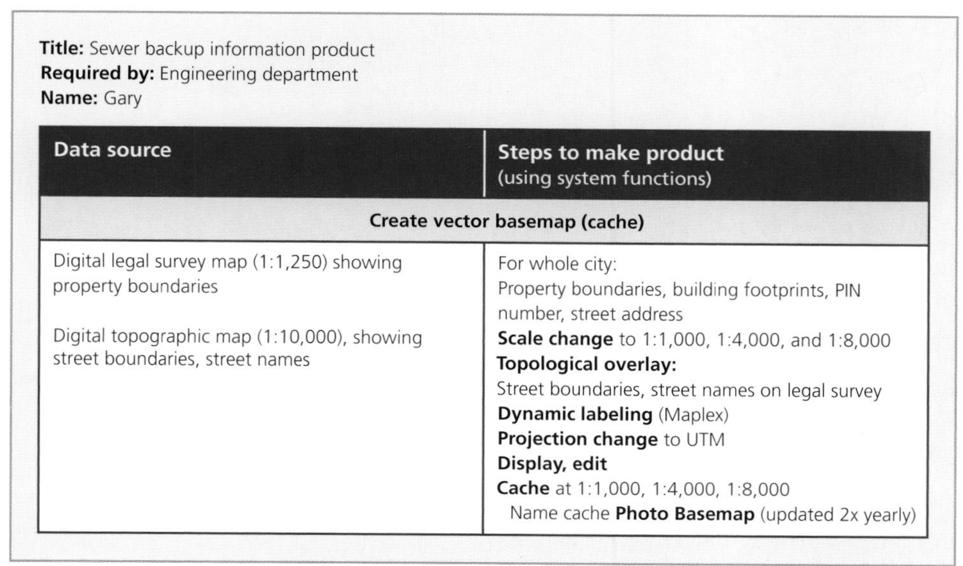

Title: Sewer backup information product
Required by: Engineering department
Name: Gary

Data source	Steps to make product (using system functions)
Create vector basemap (cache)	
Digital legal survey map (1:1,250) showing property boundaries Digital topographic map (1:10,000), showing street boundaries, street names	For whole city: Property boundaries, building footprints, PIN number, street address **Scale change** to 1:1,000, 1:4,000, and 1:8,000 **Topological overlay:** Street boundaries, street names on legal survey **Dynamic labeling** (Maplex) **Projection change** to UTM **Display, edit** **Cache** at 1:1,000, 1:4,000, 1:8,000 Name cache **Photo Basemap** (updated 2x yearly)

Figure 6.12 **Steps to make Gary's information product.**

Create raster basemaps (cache)	
Ortho imagery at 1:2,000 (NCC) City Basemap at 1:1,000 and 1:4,000	**Scale change** to 1:1,000, 1:4,000, and 1:8,000 **Projection change** to UTM **Graphic overlay** of City Basemap at 1:1,000, 1:4,000, and 1:8,000 **Symbolize, display, edit** **Cache** at 1:1,000, 1:4,000, and 1:8,000 Name cache **Photo Basemap** (updated 2x yearly)
Create operational layers	
Sewer map layers at 1:1,250 • Storm sewers • Sanitary sewers	For each service area: **Attribute search** to identify boundary and sewer network with manholes as nodes, manhole number, segments between manholes, sewer segment number, and building connections and street address. **Scale change** **Display, edit, label** 1:1,000 and 1:4,000 Name **Sewer Map**
Sewer characteristics file (SIMS) • Sewer log file • Drain card file • TV reports	**Attribute search** to identify sewer type (sanitary or storm), segment #, service area, capacity, size, grade, material, date installed, physical condition, date last inspected, date last cleared, dates of incidents, and replacement due date. Create list database **Sewer Data**
Engineering department operational reports	For each sewer service area: **Create list** of current work orders, leaks per mile, customer calls, and problems. Name list **Sewer Ops** **Create pop-up windows** By sewer service areas of list items (Sewer Ops) By sewer type (storm or sanitary), sewer segment of sewer data

Figure 6.12 (continued)

Also later on, you will make good use of this right-hand column's listing of the required system functions. In fact, now as part of the IPD, you should use the information from this column to make a simple table, like Gary's (figure 6.13), listing these necessary functions and specifically how many times each will be used to make the product.

Processing complexity

You now need to assess the processing complexity of the information product. In the function utilization table shown in figure 6.13, mark complex functions with an asterisk or some other indicator. These are the functions that require more computing resources to complete. (The lexicon at the back of the book indicates complex processes with asterisks.) For Gary's information product, there are twelve instances of complex functions (ten for base layers, and two for operational layers). This earns

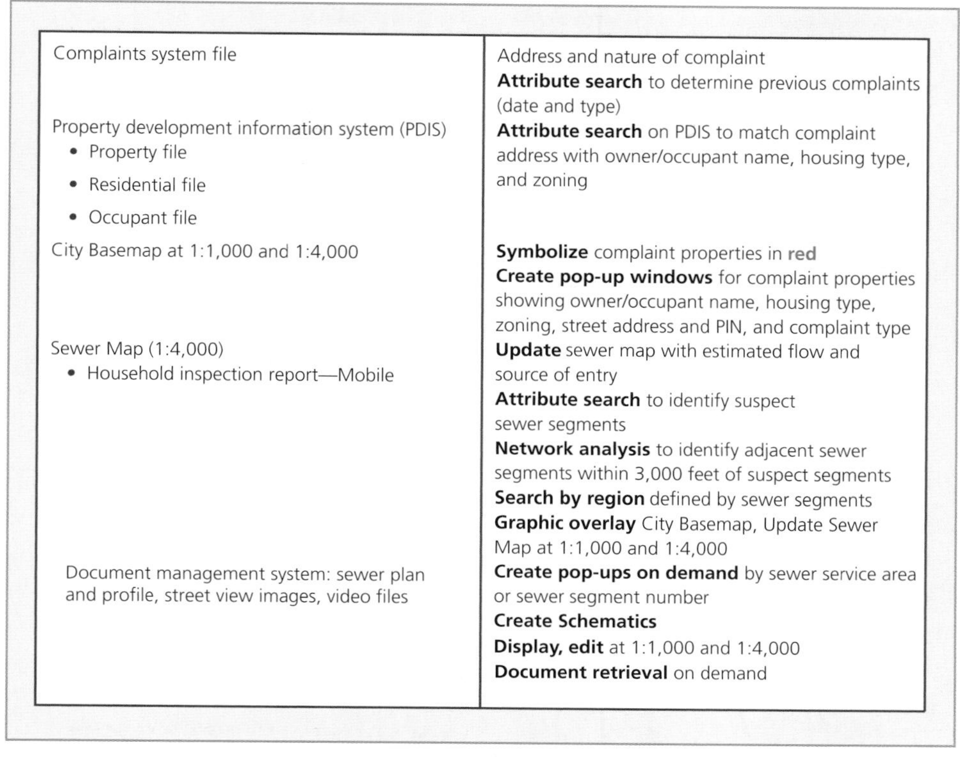

Complaints system file	Address and nature of complaint **Attribute search** to determine previous complaints (date and type)
Property development information system (PDIS) • Property file • Residential file • Occupant file	**Attribute search** on PDIS to match complaint address with owner/occupant name, housing type, and zoning
City Basemap at 1:1,000 and 1:4,000	**Symbolize** complaint properties in **red** **Create pop-up windows** for complaint properties showing owner/occupant name, housing type, zoning, street address and PIN, and complaint type
Sewer Map (1:4,000) • Household inspection report—Mobile	**Update** sewer map with estimated flow and source of entry **Attribute search** to identify suspect sewer segments **Network analysis** to identify adjacent sewer segments within 3,000 feet of suspect segments **Search by region** defined by sewer segments **Graphic overlay** City Basemap, Update Sewer Map at 1:1,000 and 1:4,000
Document management system: sewer plan and profile, street view images, video files	**Create pop-ups on demand** by sewer service area or sewer segment number **Create Schematics** **Display, edit** at 1:1,000 and 1:4,000 **Document retrieval** on demand

Figure 6.12 (continued)

Gary's information product a high overall complexity rating. (Processing complexity for active client sessions is reduced by caching the base layers.)

Display complexity

In Gary's information product, the number of layers and features are not excessively high, there are no fancy visual enhancements required for the product, and the base images will be cached, all of which would earn a light display complexity rating. However, there are two geoprocessing outputs that are displayed during the active client session: a network layer and a schematics layer. Due to these two complex layers, the information product earns a high display complexity rating.

Frequency of use

The most frequently used functions are those involved every time there are major sewer backups (those performed on the operational layers). Gary estimates that each year his department will require fifty of the information products needed to respond to basement flooding complaints. To calculate frequency measurements for each function, simply multiply the number of times each function is used per product by the number of times per year the product will be created. This annual figure will give you an idea of functional use, a primary indicator of the type of software system capable of supporting it (discussed further in chapter 9). Because basemaps are only generated a few times a year, you only need to count the operational functions at this point.

Now you can rank the functions required to generate the information product, listing the most frequent first, according to the number of times each will be used in a year to produce this information product fifty times:

1. Attribute search: 250
2. Display: 200
3. Edit: 200
4. Create pop-ups: 150
5. Document retrieval: 150
6. Graphic overlay: 100
7. Scale change: 50
8. Symbolize: 50

Title: Sewer backup information product
Required by: Engineering department
Name: Gary

Base layers	
Function	**Frequency**
Scale change	6
Topological overlay*	1
Maplex label*	1
Projection change*	2
Display	2
Edit	2
Cache*	6
Graphic overlay	3
Symbolize	1
Operational layers	
Function	**Frequency**
Attribute search	5
Scale change	1
Graphic overlay	2
Display	4
Edit	4
Label	1
Symbolize	1
Create pop-ups	3
Update	1
Network analysis*	1
Search by region	1
Schematics*	1
Document retrieval	3
*** complex function**	

Figure 6.13 **GIS function utilization table (per information product).**

9. Update: 50
10. Search by region: 50
11. Network analysis: 50
12. Label: 50
13. Schematics: 50

The function that will be used the most is attribute query. So now you know that the software Gary's organization implements must be able to perform attribute queries efficiently, at least for the benefit of this information product.

Logical linkages

Gary needs a link in the GIS database between street addresses and property boundaries in order to select the lot lines for the map. He also needs a link between a sewer segment number and the sewer network segment, the actual line in the digital database. These are key linkages between items in a list (attributes) and a graphic entity in the database (features).

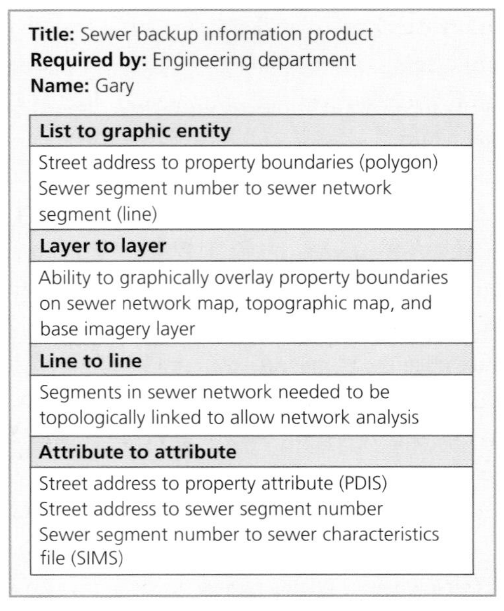

Title: Sewer backup information product
Required by: Engineering department
Name: Gary

List to graphic entity
Street address to property boundaries (polygon) Sewer segment number to sewer network segment (line)
Layer to layer
Ability to graphically overlay property boundaries on sewer network map, topographic map, and base imagery layer
Line to line
Segments in sewer network needed to be topologically linked to allow network analysis
Attribute to attribute
Street address to property attribute (PDIS) Street address to sewer segment number Sewer segment number to sewer characteristics file (SIMS)

Figure 6.14 **Logical linkages.**

Some layer-to-layer linkages are also required. Gary must be able to overlay the property boundaries on the sewer network map and the topographic and imagery basemaps, meaning they'll need to be available at a common scale and projection. Note that the degree of scale change of any dataset required to make the overlay should not be too large. Although scale changes of more than 2.5 times in either direction should be avoided generally, you may have to live with more, if the more closely projected or scaled data is not available. Heed the information systems creed: be guided by the error tolerance.

Links between graphic entities (in this case, line-to-line) are needed to establish a topological network of the sewer segments.

Finally, three attribute-to-attribute linkages must be made: the street address needs to be linked to the property attributes in the PDIS; the street address must be linked to the sewer segment numbers (to determine which segment in which sewer is backing up into each house); and the sewer segment number needs to be related to the sewer characteristics file (SIMS). It is useful for the IPD to list all the linkages in one place, as in figure 6.14.

Unfortunately, no digital citywide sewer segment map currently exists, so it is not possible to link sewer segment numbers and sewer characteristics. Without these segment numbers, it is also not possible to link street addresses to sewer segments. To the uninitiated, this may sound like a simple thing to correct: just manually determine and append the data. But in fact, the cost of creating such things can be enormous—in the millions of dollars for even a medium-sized town or city. This is because creating an overall sewer network map involves scanning old sewer plans, digitizing the real position of the manhole covers from aerial photography, and sliding the images in behind those digital points.

Fortunately, Gary's city government recognized long ago how beneficial this new database could be and decided that its development should be an infrastructure cost in the engineering department's budget, not a cost for the GIS budget. In your own organization, you may need to figure out sensible ways to get these databases built or else be prepared to include them in your GIS budget to be balanced against the benefits of the new information products.

Error tolerance

It is up to Gary to think carefully about how much error the product could tolerate and still remain useful. Is it really necessary to check every address three times to get zero errors? Even a simple operation such as rechecking an address could severely increase the cost of creating the information product. Yet error and accuracy matter because they affect the reliability of the system as a whole. To assess how much error is tolerable, Gary must ask the question, "How wrong can we be?"

Gary identifies potential errors in all four error categories (see figure 6.15). He specifies 0 percent error tolerance on referential and topological errors. This is the ideal, but if costs seem too high, he is prepared to revisit his error tolerance. Even if Gary ends up changing it, he'll derive a lot of value from examining the issue of error tolerance and discussing it with you, as the GIS planner, in specific terms. By examining acceptable error, he begins to understand the effect of errors on the information product's reliability and cost.

Wait tolerance and response tolerance

Because the sewer backup represents a time-sensitive issue, but not a life threatening one, the network wait tolerance does not apply (non-critical, which means it could be multiple hours, though it will likely take much less time with today's sophisticated hardware). Gary designates four to eight hours (one day or less) as the response tolerance, measured as the time between the first telephone complaint and availability of the complete information product at the sewer backup site.

Title: Sewer backup information product
Required by: Engineering department
Name: Gary

Type of error	Possible occurrences	Result of error	Impact on benefits	Error tolerance
Referential	Street address error	Wrong identification of property	Erroneous situation analysis	0% error
	Sewer segment number error	Wrong identification of suspect sewer segment	Wasted time while errors are resolved	
Topological	Link between property and sewer not established	Property not included in analysis	Incomplete situation analysis	0% (complete topology required)
	Break in topology of sewer network	Adjacent sewer segment would not be included in analysis	Potential source of problem may be missed	Links between property and sewer are particularly important
Relative	Location of sewer line within street	Wrong site for excavation	Increase in on-site cost	± 1m
Absolute	Graphic overlay of property boundaries on sewer network may be misaligned	Uncertainty in sewer line position as it relates to complaint property	Potential increase in on-site investigation costs	± 1m

Figure 6.15 **Error tolerance table.**

Current costs

Gary estimates that using current non-GIS methods takes one hundred hours to create this information product, or $6,000 in labor costs each time the product is created (figure 6.16). In addition to labor costs, he accounts for another $100 in material costs, for a total cost per information product of $6,100 right now. Creating this information product fifty times per year—without the GIS—costs $305,000 annually. That is the current cost of creating the product within an operational budget for sewer maintenance of $12 million per year.

Title: Sewer backup information product
Required by: Engineering department
Name: Gary

To create product	Hours	Cost (U.S.)
Labor **Professional** **Technical**	100	$6,000.00
Materials		$100.00
Total cost		$6,100.00

Figure 6.16 **Current cost without GIS.**

Benefits analysis

The GIS could create the same product in four to eight hours, compared to one hundred hours of manual effort, reducing the time involved by more than 90 percent. Some staff members would remain involved, but still, staff time savings of more than 80 percent would be realistic. This represents a real cash savings.

In addition, with GIS creating it faster, the product could be available within a day, as opposed to three weeks later. This is a direct benefit to the organization, helping them react more quickly to a flooding situation. With this faster reaction time, perhaps they could solve the flooding problems quickly enough to keep homeowners from becoming angry, and minimize liability. Improved timing may also make it possible to coordinate repairs with maintenance operations already under way, another agency benefit. Gary can see that it's entirely reasonable to expect benefits of at least $120,000 per year.

One more direct benefit to the agency will be a reduction in liability. Long delays in solving backup problems have already led to a backlog of ten court cases involving unresolved sewer backup problems, including a class action suit for $700,000 by a group of homeowners. The cost of legal preparation by engineering department staff alone could be reduced by an estimated $60,000 per year.

Future and external benefits include improvements to the environment as a result of resolving the flooding and

Title: Sewer backup information product
Required by: Engineering department
Name: Gary

Savings

Current data compilation is time-consuming—100+ person hours for each flooding (fifty per year). This time could be reduced by 90 percent. Staff time savings of 80 percent per year on current workload.

Benefit—$120,000/year

Benefit to organization

Timing of product output will be improved significantly. Information provided on basement floodings just after the storm occurs.
Solutions may be applied in time to verify the correction during the rainy season.

The improved timing will make it possible to coordinate repairs with maintenance contracts currently under way, which would save considerable sums.

Reduced liability. There is currently a backlog of ten court cases awaiting trial. The number of court cases increases as time goes on. Staff spends considerable time collecting data for court cases. Legal costs would be reduced.

Benefit—$60,000/year

Future and external benefits

By resolving basement-flooding problems, the environment is improved. To relieve basement floodings, sanitary sewers are pumped into the storm sewers, and there is an increase in pollution levels in freshwater streams and lakes. Sewage treatment costs would be reduced.

Benefit—$12,000/year

Gary Jones *Gary Jones*

Department head ________

Figure 6.17 **Signed benefits page: end of IPD.**

sewage backup problems more quickly. Currently, when the sewers back up, their flow is diverted to the storm sewers, which normally run into the lakes and streams. Some of the polluted storm sewers are diverted to the local waste processing plant, increasing its volume substantially. The costs of extra processing could be lowered by $12,000 per year.

As you can see on the last page of Gary's IPD (figure 6.17), these benefits could result in savings of $192,000 per year. Expanded over ten years, the benefits would amount to $1.9 million, without even taking inflation into account. These figures strengthen the case for building the new databases. You can now compare the benefits that will result from the new information against the costs of data acquisition and system implementation. Management accountability framework objective #12—improve average problem resolution time of basement flooding by 20 percent in the next three years (see box on page 59)—will be achieved by using GIS.

Gary's department head recognizes the benefits of GIS from this comparison, too, and indicates his approval of Gary's completed IPD by signing off on the benefits pages. With Gary initialing every page, then putting his signature on the last page, his IPD is ready to be handed off officially to the GIS team.

The sample forms used in this case study have facilitated the successful collection of IPDs in state, regional, and municipal organizations over many years by Tomlinson Associates Ltd. A template IPD form can be found on the supplemental DVD. Examining this template will help you complete the information needed to ensure that your requirements are fully evaluated.

Master input data list (MIDL)

In creating the IPDs, you identified the data needed. Now you're going to use this knowledge to create a new document: the master input data list (MIDL). The MIDL is a detailed list of all the source datasets that need to be entered into the GIS system to generate all the new information products. The MIDL should identify each source dataset (with its name, ID number, and provider) and include assessments of the data volume (amount), format, availability, and cost.

The MIDL

- is the master list of data that needs to go into your system; and
- identifies the work that will be involved in putting the data into your system.

The MIDL will guide the effort of setting up your database when the time comes, so estimate as closely as you can how much work will be required, such as the amount of digitizing and alphanumeric data input, along with how much data must be available each year.

To be included in the MIDL, each dataset must be needed and specified for at least one information product. No other data should be described in the MIDL. If someone comes to you offering "good" data to include in the GIS, ask that person, "Which information product needs this data?" Even so-called good data should not make it into the MIDL unless it is necessary for the creation of at least one information product. Having this business rule in place forces people to think in terms of the end products from the GIS; without this rule, your GIS data directories would quickly grow into an array of layers and features that nobody understands because they don't ever get used.

A member of the GIS planning team should be responsible for creating the MIDL; it's perfectly appropriate for the team leader to undertake this task. Whoever is responsible should bring a working knowledge of the characteristics of data, both the data at hand in the organization and that coming from elsewhere. If GIS planning becomes your career, with experience you'll learn to create the MIDL at the same time you create IPDs. This way, as you identify the datasets necessary to make information products, you can record the dataset names and characteristics directly onto the MIDL as you go (see "Focus: Data shoe box" on page 75).

Components of an MIDL

A master input data list includes the following four components:

1. Data identification details
2. Data volume considerations
3. Data characteristics
4. Data availability and cost

Data identification details

Data identification details distinguish each dataset using a name or number. Because different people or departments may have different names for the same data, it is important to create a common identification name or number that can become the standard throughout the organization. You should also note if there is metadata (and its completeness), as well as the source (the agency name and location or web URL).

Data volume considerations

The volume or amount of data you need will affect your system design. Spatial data tends to demand a lot of disk space; having adequate, scalable storage will save you grief down the road. If you're lucky, some of the data will come spatially referenced in the format and scale and projection you need it. But realistically, much data will require conversion from another format. Some will not even be digital data; these nondigital datasets will require conversion to a GIS-readable format. The amount you'll need to store, as well as the amount you'll have to preprocess, affects your working strategy for data storage and handling. You should evaluate the number of datasets you need overall, the source data medium, the primary record type, the primary record volume, and what percentage is digital now. Also consider the quantity and size of their attribute files. Will you need two datasets or two hundred? Are the attribute files associated with these datasets large or small?

Data characteristics

Understanding the format and pedigree of the data you intend to incorporate into the GIS will guide your database design and, ultimately, the selection of software and hardware. You need to know whether the data is available as digital files already in an appropriate GIS format: is the format one you can use directly in your system or do you need to convert the data from another format before using it? What are the data characteristics that will determine your input method (scanning, digitizing, or text input)? Note the detailed breakdown in figure 6.18 even calls for counting the number of keystrokes if applicable; in other words, the more specifics the better.

For hard-copy data, determine how best to convert to digital—will there be scanning, digitizing, complex label creation? If there is associated attribute data, can it be imported into the GIS or will there be significant data entry involved? For digital data, note the data type, file format, and map projection.

As you consider data, you need to decide what data can be cached in order to preprocess as much as possible and reduce display processing time. Caching is extremely desirable if you plan to serve maps on the Internet. A cache stores preprocessed data, such as base layers for a map or the output of commonly used geoprocessing functions. Cached data has the benefits of very fast retrieval and reduced platform loads.

You can read a more detailed description about various data characteristics and capabilities in chapter 7, "Consider the data design."

Data availability and cost

Later on as you develop your implementation strategy, facts about the availability of data and its cost become imperative. Sometimes, it is cheaper and easier to re-create data than to edit and update existing commercial or government data. Some of the questions you need to answer are the following:

- Does the data exist in a digital format or will you need to convert it?
- What is the cost of acquisition? Do royalties need to be paid?
- Is it possible to partner with another agency to share the cost of data?
- What is the cost of converting data from one format to another?
- Who is going to be responsible for updating and maintaining data?
- Are there any restrictions on the use of data?

Assembling the MIDL

The components of the MIDL (figure 6.18) summarize the information to be collected for each dataset. You will fashion your own according to what is appropriate for your project.

The master input data list details the basics: data name, source, volume, the outright cash cost, and any significant variables (such as conversion and availability schedules); the specific items of information will vary somewhat depending on the project.

Capture everything that will affect your acquisition and use of the data you need. Data about the data (metadata) doesn't use much storage, and taking the time to set it up correctly will pay dividends down the road when stakeholders start questioning the basis of your analyses. As a general rule, it's a good idea for every project to gather as much metadata as possible. (Your GIS system should have built-in metadata handling capabilities.)

Any book on IT planning would be remiss not to drag out the shopworn phrase "garbage in, garbage out" to drive home the importance of stocking your GIS with only the most reliable and accurate data. This is not to say that lower resolution or more generalized spatial data cannot be loaded into a GIS; just be sure that the associated metadata reflects its limitations.

Functions needed to input data

Once you've collected the information for the MIDL, you can evaluate the basic system capability functions

Component	Details needed	Notes
1. Data identification	Dataset name	
	Dataset number	
	Source agency name	
	Internet location (URL)	
	Metadata available	Yes or no
2. Data volume considerations	Source data medium	
	Digital data format	
	Percentage available now in digital format	
	Primary record type	Line, sheet, etc.
	Primary record volume	Number of primary records
	Total data volume	

Figure 6.18 **Components of the MIDL.** (continued on next page)

necessary to put each dataset into the GIS. For every dataset in the MIDL, list the functions required to put the data into the database. These data-handling functions will differ depending on the type and characteristics of the data to be input. For example, it's a different set of operations altogether for the computer to input a hard-copy map than to input a digital image. Be well informed about potential pitfalls—you need to make sure you understand about data interoperability and data conversion (more on this in chapter 7).

For now, set aside this list of functions required to input the data. Later on, you will combine it with the list of functions required to make each information product, identified in the IPDs. Eventually, in combining your two

Component	Details needed	Notes
3. Data characteristics		
Digital data considerations	Data type	E.g., digitized maps, text, photogrammetry, survey measurements, images, GPS
	File format	E.g., shapefile coverages, file geodatabase, OGC-GML, raster, DXF
	Map projection	
	Dynamic or static	
Hard-copy data: Scanning considerations	Sheet size	In inches or centimeters, typical and maximum
	Minimum scan resolution	In dpi, without compression
	Legibility	Percentage of total data volume in high, medium, low, and difficult categories
Hard-copy data: Graphic portion	Size	In inches or centimeters, typical and maximum
	Schematic	Yes or no
	Photo image	Yes or no
	Map projection/datum	
	Measurement (COGO) data volume	Typical and maximum volume or coordinates Typical and maximum number and size of observations
Hard-copy data: Digitizing effort	Polygons per sheet	Typical and maximum
	Lines per sheet	In inches or centimeters, typical and maximum
	Points per sheet	Typical and maximum
Hard-copy data: Text portion	Lines per sheet	Typical and maximum
	Data elements per record	Number of fields per record, typical and maximum
	Total alphanumerics per sheet on input	Number of keystrokes, typical and maximum
4. Data availability and cost	Percent coverage available now	Or date when it will be available
	Currency	Date of data capture and date of most recent update
	Restrictions on use	
	Cost of dataset acquisition	
	Royalties	On acquisition and use

Figure 6.18 (continued) **Components of the MIDL.**

FOCUS

Using and creating a data cache

Consider creating a cache for reusable basemap data. Map caching is the process of pregenerating large collections of tiled map images for rapid display of large or complex maps. Despite increasingly fast dynamic map drawing, caching is still necessary for efficient, fast mapping applications, essential for maps served on the web. Because the cached layers are not dynamically rendered, display rendering time is significantly decreased; once the cache has been downloaded, the client doesn't have to wait for it to reload with every new request. Dynamic (operational) layers can be mashed with fully cached base layers with high performance.

You may choose to cache an entire basemap, or you may generate an initial partial cache based on areas of interest or different levels of resolution. Remaining areas can be cached on demand. On-demand caching allows the client to create cached files on the fly; also, if a cached map is edited, the new map is automatically updated with the next user map request.

The latest software advances offer extremely compact cache storage. The compact cache groups tiles into large bundles (128 × 128 = 16,384 tiles) for streamlined storage and easy transfer from one machine to another. You can also create mixed-mode caches, where there are different file formats in the same bundle. The mixed-mode cache supports transparency where needed in image caches; for example, you might use JPEG files (highly compact) except where overlapping of cache layers occurs, where you would use PNG32 files (which take up more storage space and bandwidth) for transparency. This is an enormous space saver for large caches.

For the organizations that have multiple users or want to collaborate on building a cache, import and export tools exist that can be used for a full or partial cache. Portions of the cache can be disconnected for editing, then checked back into the parent dataset seamlessly. The software treats the cache as a raster dataset, so working with even large caches has become quite simple.

There are some drawbacks to caching. The initial data cache creation can be a large time investment, as every possible display extent and resolution must be pregenerated. For example, consider caching a basemap of the city of Ottawa (population 1 million) at a scale of 1:2,000. In this scenario, hundreds of thousands of tiles will be produced, requiring 1–2 days of work. There is limited functionality of cached data; it is designed to be static, with minimal alterations. Overlay of more than three cached multilayer displays may be slower than generating dynamic noncached displays. Finally, keep in mind the cache is separate from the maintenance data source, so it requires additional dedicated storage space. Cached data also needs to be updated separately. Despite these limitations, if you are publishing a high volume of web mapping services, you will probably find the limitations to be outweighed by the benefits of caching. Caching options are offered when publishing web maps.

Google Earth is a good example of a vast collection of cached maps. The first time you view a particular tile or set of tiles, it may take some time, but then they are saved in your Internet browser's cache, for quick subsequent display.

FOCUS

Data shoe box

Because some of the information required for the master input data list (MIDL) is detailed and requires time to assemble, it is a good idea to begin collecting information for this document while preparing the IPDs. You can use the data inventory form shown in figure 6.19 to enter information about datasets identified during the IPD building process. Don't get hung up trying to answer all the questions right away. The idea is to give each dataset a name and a number and describe the data well enough to recognize it when you return to it. You can fill in the other portions of the form later. As you can see, this form covers most everything you can think of to describe data, including dataset name, source, scale, projection, format, description, and volume.

It's good practice to attach sample data to the form. This might include representative maps, with a copy of the legend, or all the columns for two to three records of tabular data (lists). Such samples serve to clarify the functions that the user must perform to create or use the data. These data samples, with inventory forms attached, should be collected together in a physical place, what we call a *data shoe box*. If data is not represented with a sample in the shoe box, it is considered nonexistent in terms of the GIS. Many people new to GIS are unclear about what data exists in their organization. The shoe box is the test! Use a separate data inventory form for each dataset, staple a sample of the data to it, and put it all in the shoe box.

Data inventory	
Interviewer:	Date:
Name:	Department:
Phone#:	Division/Office:
Dataset name:	Dataset number:
Source agency:	Metadata? (circle) Yes No
Internet location (URL):	
Data type: (circle) Hard-copy map Digital map Text data file Air photo Digital data file Text Imagery GPS coordinates Survey measurements	
Other:	Source data medium:
Digital data format:	
Percentage available now in a digital format:	
Total data volume (# of pages, CDs, maps, etc.):	
Scale of source data:	Map projection/datum:
Photo image? (circle) Yes No	
Digitizing effort (per sheet):	# of polygons:
	# of lines:
	# of points:
Date of data capture:	Date of last update:
Percent of coverage available now:	
Date entire coverage is available:	
Restrictions on use:	
Cost of dataset acquisition: $	
Royalties on acquisition and use: $	

Figure 6.19 **Data inventory form.**

lists—the type of function and number of times the function will be used (from the MIDL and IPDs)—you will have the complete list of software functionality your system requires. That will be the piece of paper software vendors will need to see in order to compete for your business.

Setting priorities

At the technology seminar, you began to get a sense of which information products were more important than others to your organization's workflows. Now you must clarify those priorities because, of course, all the information products required from a GIS cannot be created at once. You need to prioritize your list according to the relative importance of the contribution of each product to your organization's objectives. These priorities will be ranked using a scoring method, a group-consensus method, or some combination of the two. Remember, this ranking is the thing that will guide which information products get delivered first, so make sure the process you use incorporates the input of upper management.

When assigning priorities, you must rank information products in strict numerical order—no two can have the same rank. For your process of prioritizing, choose either (or some combination of) the scoring method or the group-consensus method.

The scoring method

In the scoring method, the GIS team leader creates a simple model by which each information product is scored based on the benefit it will provide: ease of production, relevance to the organization's strategic plan, facilitation of specific management accountability framework objectives, whatever. The planning team members decide the appropriate criteria for scoring. The GIS team leader can allocate scores alone initially or in consultation with the rest of the team throughout the process. Once the team is satisfied, the list of information products and their priority scores—rankedin order of the highest first—is submitted to executive decision makers for comment and approval. Senior management should and will make the final decision on priorities.

The group-consensus method

The group-consensus method is a less structured approach to assigning priorities: you get all the managers and decision makers into a room together and work until consensus is reached on the priorities assigned to all the information products. This is not always an easy process; in fact it can be quite challenging. In such a situation, you can choose between the "snake in the grass" approach, where the process is strategically molded and guided by managers, or the "blood on the floor" approach, where nobody leaves the room until the priorities are resolved (the latter being the more painful!).

External consultants are not welcome in that room; they don't have enough of a vested interest in the organization's missions and objectives to be able to effectively prioritize the information products. It's management's concern where to place priorities, taking into account the opinions of those who will create and use the product. As conditions change in the future, rankings may need to be adjusted. This initial assignment of priorities, however, gives your GIS implementation essential direction and validity.

Input data priorities

In the same way that you ranked the information products by priority, you should rank the input datasets to determine priority for acquisition or creation. Reorder the MIDL based on the order in which data is required to make the information products, starting with the highest-priority information products. On the new version of the list, datasets to be used first will have the highest priority. The end product of this simple step should be the assignment of data priorities to each dataset in the MIDL. In the list itself, the highest priority (most immediately needed) data will be on top.

Chapter 7

Consider the data design

The data landscape has changed dramatically with the ubiquity of the Internet and the proliferation of commercial datasets. Developing a systematic procedure for intelligently navigating this landscape is critical and requires an understanding of spatial data characteristics.

Now that you know what data you need to make your information products and what it will take to enter it all into your database, perhaps a conception of the infrastructure required to support it is beginning to take shape in your mind. But before you get too far in thinking about your system design as a whole, you'll need to think about how to design the system for data in more detail. The data design comes first because the characteristics of your data determine, in large part, what you'll need from the system architecture.

The issues of data design (in this chapter) and the choice of a database model (in the next chapter) both boil down to two concerns: First, will the data represent the real world in a way that is useful to you? For example, will it let you describe features, details, behaviors, conditions, and anything else that you need to use in your calculations and analysis? Second, does the use of a particular database model inhibit or facilitate some of the things you want to do? Does it prohibit the use of some types of data? Is it very complex and slow to execute? Is it scalable? We'll explore database modeling in the next chapter, after first focusing on what you need to know about your data.

The information contained in this chapter will help you refine your information product descriptions and the MIDL, and get you up to speed on the factors that will affect the system design, which occurs during implementation.

Data characteristics

Part of developing a systematic procedure for creating the data design is having a thorough understanding of the characteristics of your data, whether you are creating or acquiring it. These characteristics include each dataset's scale, resolution, map projection, error tolerance, and how the data affects the intended information products. Sometimes you'll need to create multiple versions of the same datasets at different scales or resolutions to generate the map output specified by the IPD. These will all be stored as individual layers in your database.

Scale

Scale is the relationship between the distance on a map and the corresponding distance in the real world. If a map has a scale of 1:24,000, then one map unit (one inch, for example) on the map is equal to 24,000 map units (e.g., 24,000 inches, or 2,000 feet) on the ground. Map scale can also be expressed as a statement; for example, one inch = two thousand feet. The scale of data reflects resolution and its relative accuracy on a map: the larger the scale, the more detailed the dataset shown. Map scale numbers are counterintuitive: the lower the number, the higher the resolution (and the larger the scale). So 1:6,000 data is significantly higher resolution than 1:100,000 data.

Features drawn in large scale (e.g., 1:6,000) show greater map detail because they are more closely representing the real-world features. On a small-scale map (e.g., 1:100,000) the features are usually generalized or aggregated. Both are useful and serve a purpose.

Consider figure 7.1. A small-scale dataset shows the river on the map as a single, gently curving line. A large-scale dataset shows a smaller part of the river as a polygon depicting the banks and the width of the river.

What scale or scales to set up in your GIS database (i.e., how high a resolution of data to acquire) is a critical factor in the system design. If the scale of primary data is too large, the data volume could overwhelm the computational processing resources. This becomes especially significant when you're trying to build quick-response systems. If the scale is too small, the database might not contain the details needed, and so would not be able to perform the best analysis or reliably provide the specified information products. The "native" format of your source data (the format in which it comes to you), as well as your budget for computing power, will advise this process.

Once you've chosen the appropriate scale for your GIS database, you may need to convert some data to this common scale, either in advance or in real time, as one of the steps required to create information products to fit with other data.

A good standard is to avoid changing scale by more than two and a half times in either direction. As in figure 7.2, if you start with a 1:50,000 scale dataset and reduce the scale two and a half times (multiply 50,000 times 2.5), your scale will be 1:125,000. Thus, if you create information products from source data that has a scale smaller than 1:125,000, you may run into data profusion and legibility problems. If you start at 1:50,000 and go two and a half times larger (divide 50,000 by 2.5), your maximum scale will be 1:20,000.

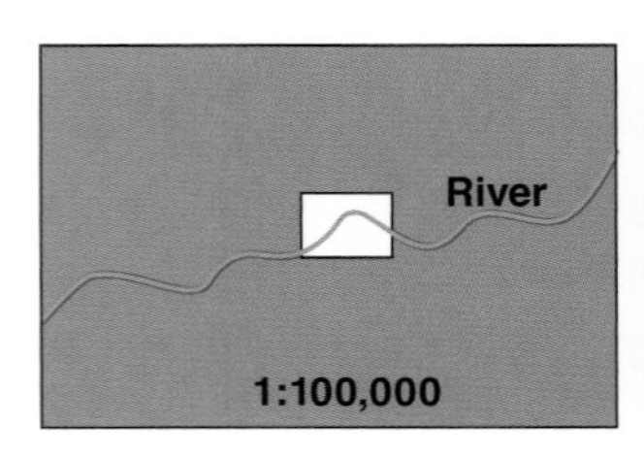

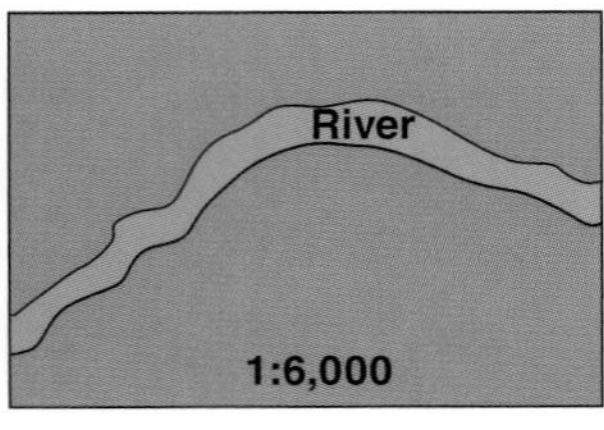

Figure 7.1 **Dataset in small scale (left) and in large scale (right).**

In some cases, you may need to store source data at more than one scale in the database, such as when applications will be performed at both small and large scales. Sometimes you may want to store basemap data acquired from external sources at the native scale in which it was received. Although a GIS does not typically store the scale of the source data as an attribute of a dataset, scale is a useful indicator of accuracy, which is why you note scale in the IPD and the metadata. All users of the data must understand that true spatial analysis—such as when you're using geographic overlay functions to derive new information—is only as accurate as its lowest resolution layer. If more than one scale is represented in the database, it should be well-documented in the metadata. Note that if the different scales are from different sources, they may not agree.

Scale influences the cost as well as the accuracy of the resulting database. For example, the number of map sheets needed to cover the same area in your database increases exponentially with the scale. Therefore, mapping at 1:6,000 is sixteen times more expensive than mapping at 1:24,000. As the number of sheets needed for your application increases, so does the cost. And because scale does introduce such a significant effect on cost, you must take a close look at the actual needs as specified in the IPDs. Don't build a rocket to Mars if all you need is one to the moon. The purpose of the application will determine the appropriate scale.

Resolution

Consider resolution defined here as the size of the smallest features that can be mapped or sampled at a given scale. The resolution of a map is directly related to its scale. As map scale decreases, resolution diminishes and feature boundaries must be smoothed, simplified, or not shown at all. There will be a minimum polygon size and line length you can represent at a given scale. Features below these resolutions are merged into surrounding data, converted to a point, or deleted. In figure 7.3, you can see that a two-acre polygon is visible at a large scale of 1:24,000, but the same polygon appears as a point at the small scale of 1:500,000. Resolution also determines the distance between sample points in a grid or lattice format (for example, satellite imagery—see figure 7.4).

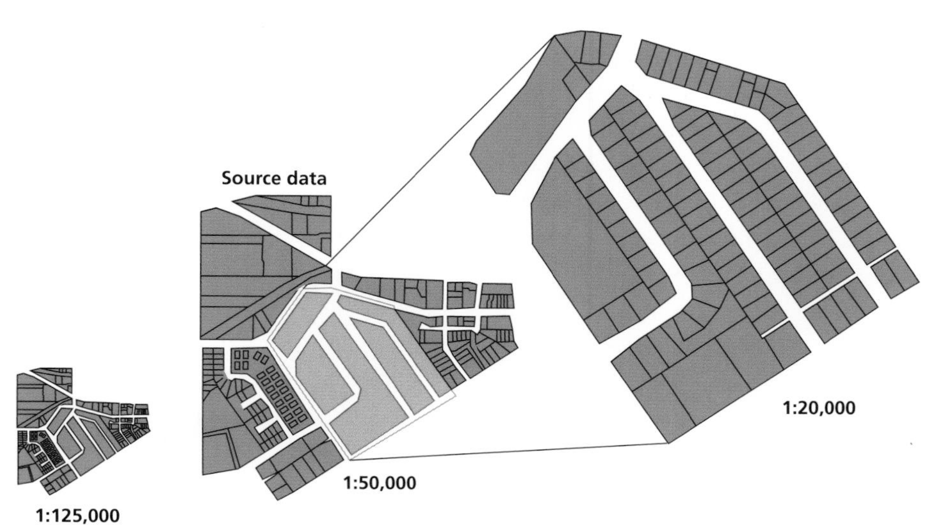

Figure 7.2 **Minimum (left) and maximum (right) scale.**

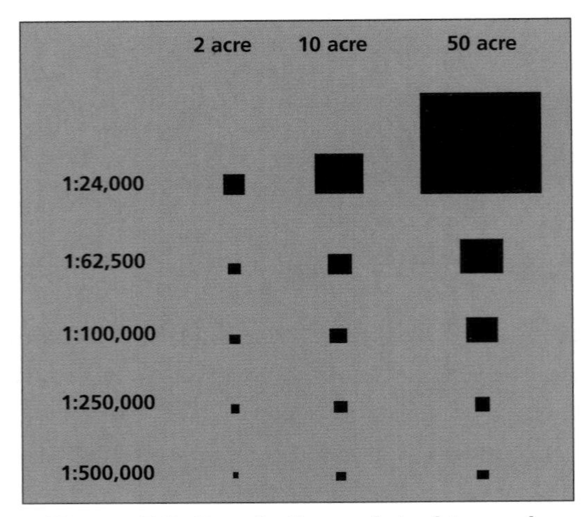

Figure 7.3 **Resolution related to scale.**

Like scale, your data must have the minimum resolution necessary to create satisfactory information products. You do not necessarily need the highest resolution available. For example, a city's land-parcel data must be high resolution, but a web application showing interstate travel routes would have to be small scale. Keep in mind that resolution also contributes to data error. Your information product descriptions include error tolerance and will help you determine the required resolution.

Map projection

The choice of a map projection is another crucial step in the GIS database design process. A basic understanding of map projections is important in order to determine the one best for your needs.

Paper maps are perhaps the most common source of geographical data in the world. Because the planet Earth is spherical and maps are flat, getting information from the curved surface to the flat one requires a special mathematical formula called a *map projection.* A map projection simply converts Earth's three-dimensional surface to the flatlands of paper—a decidedly two-dimensional space. The process of flattening the Earth creates map distortions in distance, area, shape, or direction. The result is that all flat maps have some degree of spatial distortion.

The type of projection used determines the degree and type of distortion found on the map. A specific map projection can preserve one property at the expense of the others, or it can compromise several properties with reduced accuracy. You should pick the distortion that is least detrimental to your database when selecting a projection type. Today's GIS programs offer an array of projection choices, but read the supporting documentation and "help" files before making any database decisions or attempting to reproject any data.

Datums are another important map aspect related to projection. A datum provides a base reference for measuring locations on Earth's surface. It defines the origin and orientation of latitude and longitude lines and assumes a shape for the globe on which they fit. The most recently developed datum is the World Geodetic System of 1984 (WGS84). It serves as the framework for satellite-based location finding worldwide (i.e., the Global Positioning System—GPS).

There are also local datums that provide a frame of reference for measuring locations in a particular area. The North American Datum of 1927 (NAD27), the North

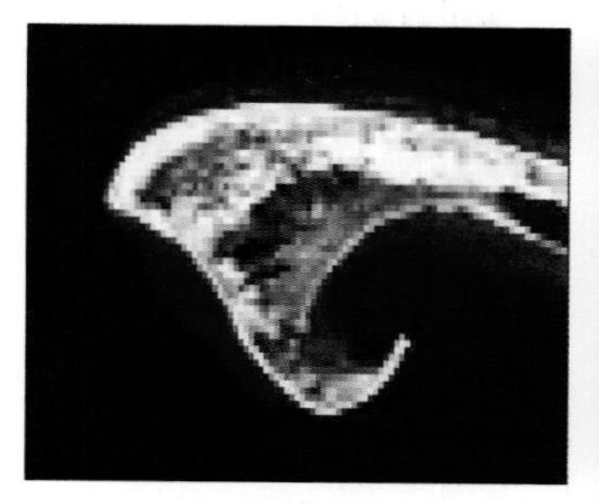
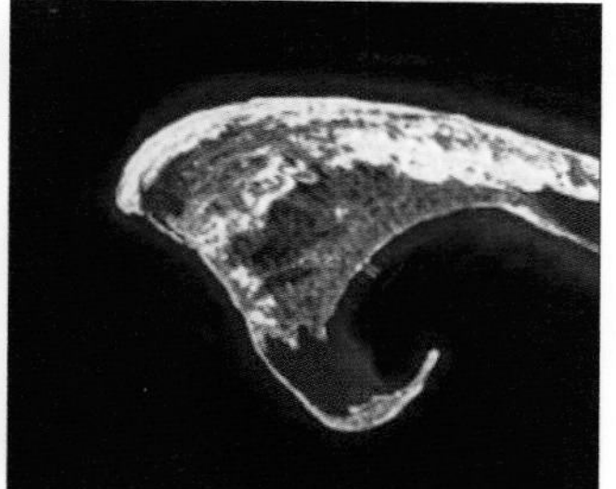

Figure 7.4 **The image on the left has a low resolution of 150 meters, where each pixel represents 150 real-world meters. The image on the right has a high resolution, 1-meter, where each pixel represents 1 meter in the real world, resulting in a much clearer image.**
Courtesy of Earth Satellite Corporation and i-cubed (`http://www.i3.com`).

American Datum of 1983 (NAD83), and the European Datum of 1950 (ED50) are local datums in wide use. As the names suggest, NAD27 and NAD83 are designed for best fit in North America, while ED50 is intended for Europe. A local datum is not suited for use outside the area for which it was designed. Maps of the same area but using different datums (e.g., NAD27 and NAD83) may result in the same features showing up in different locations on the map—clearly an unacceptable result.

To effectively use spatial data from a map, you need to know the projection of that map and the datum. Unfortunately, some people put out data without defining the projection. Often, input maps are in different projections, requiring a conversion or coordinate transformation. Your GIS software should support changing your data's projection and datum.

The amount of data distortion you'll experience due to map projection is related to scale. The larger the geographic area covered by the map (the smaller the scale), the more distortion from projection you will experience. When you see the actual scale varying at different points on a map, that is the result of map projection, and you can use it as a guideline for distortion.

FOCUS

Data accuracy

The accuracy of a dataset depends on the scale at which it was created, the resolution required for meaningful analysis, and the determined error tolerance. Data accuracy ensures that the layers in your database will serve as source data for meaningful analysis. For example, think of what would happen if you forgot that the resolution of a certain input dataset was only 1:200,000, while all the other datasets in that map library were digitized at a much higher resolution of 1:6,000. In an overlay analysis, using the lower resolution dataset with the rest of the data could profoundly compromise the validity of the resulting output. Worse yet, users unaware of the inaccuracy might base a crucial business decision on erroneous analysis. Making sure this can't happen is incumbent on the GIS leader, and the way to make sure is through rigorous documentation in the metadata.

Error tolerance

Understanding the types of error you may encounter in developing and using your GIS is very important but frequently overlooked in the GIS planning process. Error, because it is related to resolution and scale, is also directly linked to cost: reducing error costs money. Everyone involved in designing and using the GIS must have a sound understanding of what error means and how much is acceptable or unacceptable. This threshold will cost a certain amount. Some error is tolerable as long as the usefulness of the information product is maintained.

When people in your organization request an information product, they rightfully expect to receive benefit from using it. They expect it to save staff time, increase the organization's effectiveness, or contribute new value to the enterprise. The product's benefit is lost if there is so much error that it isn't useful. Worse still, the product could become an extra cost if someone makes a wrong decision based on error-ridden data.

As laid out in chapter 6, the four types of error are referential, topological, relative, and absolute. To review, *referential* error denotes error in label identification or reference. For example, in the case of a sewer-management application, are the correct street addresses on the correct houses? Are the correct sewer segment numbers on the correct sewer segments? *Topological* error occurs when there is a break in a needed linkage, for example, where polygons are not closed or sewer networks are not connected. *Relative* error is inaccuracy in the positioning of two objects relative to one another. For example, in a sewer-management application, imagine the roadway is thirty feet wide. The sewer maintenance crew needs to know where to excavate. It is important to know, relative to the property and the roadway, where the sewer is located. *Absolute* error, on the other hand, refers to the misidentification of the true position of something in the world. Absolute error becomes an issue when you are bringing different maps from different sources together in a graphical or topological overlay or combining GPS reading and maps.

Error can affect many different characteristics of the information stored in a database. Map resolution, and hence the positional accuracy of something, applies to both horizontal and vertical dimensions. The positional accuracy depends on the scale at which a map was created.

Typically, maps are accurate to roughly one line width or 0.5 millimeters. For example, a perfectly drawn 1:3,000 map could be positionally accurate down to 1.5 meters, while a perfectly drawn 1:100,000 map could only be positionally accurate down to 50 meters.

Figure 7.5 shows the relationships at play when determining a suitable map scale. Consider the minimum area to be measured and the amount of tolerable error. The expected percentage error in area measurements is derived from the minimum area mapped and the map's scale.

In the error tables from the short case study that follows, you can see further how map scale, the minimum size area you need to map, and error tolerance are related to one another. As you are developing your design for data, keep these relationships in mind.

For example, if you decide to map your entire database at 1:24,000 (because that is what you have readily available, and twenty of your thirty information product requests require this level of accuracy), you will end up creating several information products with little or no value. This will result in unmet user expectations and perhaps discredit the entire GIS. It is important to understand error and be able to effectively communicate its consequences to the people in your organization.

Given information	Resulting information
Minimum area, percentage error in area	Map scale
Minimum area, map scale	Percentage error in area

Figure 7.5 **Error table.**

Case study: Determining the required positional accuracy

The concept of error tables is confusing for many people, so let's go over it again using a simple example. Marcella works at a city's economic development department attempting to lure businesses in the warehousing and distribution industry to locate in her city. These businesses typically require good rail and highway access plus relatively level land parcels of twenty-five acres or more. Marcella needs a map that identifies these types of sites with plus or minus 5 percent error in area. (If she provides inaccurate information to new businesses, she won't get very far.) Marcella has requested that map as her information product.

If you were in the role of planning the data design for the city's GIS, you would need to review her information product description and determine the minimum scale for mapping parcels in order to fulfill Marcella's request. She can determine the minimum scale—and you can check her IPD—by using the two types of error tables that follow (figures 7.6 and 7.7):

- Map scale for a given area and error tolerance—shows the scale at which the map must be created given the minimum area that must be measured and the percentage error that is tolerable.
- Percent error in area measurement for a given area and map scale—shows the percentage error in area measurement that can be expected as a result of the minimum area being mapped and the map scale.

These tables are used to show the resulting scale or percentage error if values for minimum area and its incumbent information (figure 7.5) are given. They assume 0.5 mm positional accuracy of mapping and an average case distribution of error.

Map scale for given area and error tolerance					
Minimum Area	% error in area measurement				
Hectares (ha)	1%	3%	5%	8%	10%
0.01	95	286	476	762	952
0.1	301	904	1,506	2,409	3,012
1	952	2,857	4,762	7,619	9,524
10	3,012	9,035	15,058	24,094	30,117
100	9,524	28,571	47,619	76,190	95,238
1,000	30,117	90,351	150,585	240,935	301,169
1 hectare (ha) = 10,000 m^2 = 2.471 acres					

Figure 7.6 **Determining appropriate scale.**

Percent error in area measurement for a given area and map scale					
Minimum Area	Map scale				
Hectares (ha)	1:1,000	1:5,000	1:10,000	1:15,000	1:50,000
0.01	10.5%	52.5%			
0.1	3.3%	16.6%	33.2%	49.8%	
1	1.1%	5.3%	10.5%	15.8%	52.5%
10		1.7%	3.3%	5.0%	16.6%
100			1.1%	1.6%	5.3%
1,000					1.7%
1 hectare (ha) = 10,000 m^2 = 2.471 acres					

Figure 7.7 **Expected error.**

First, Marcella must convert her minimum site size from acres to hectares. There are 2.471 acres in a hectare, so 25 acres is approximately 10 hectares. To determine the appropriate scale, she will use the map scale for a given area and error tolerance table (figure 7.6).

She finds her minimum area in the first column of the table. Next, she finds her error tolerance (5 percent) along the top row of the table. The intersection of 10 hectares and 5 percent error is 15,058. The minimum map scale needed to create Marcella's information product is therefore rounded to 1:15,000.

Checking over her calculations using the table in figure 7.7, you find that by using maps at a scale of 1:15,000 or larger, Marcella's application will deliver the required positional accuracy.

Test your understanding of this concept by completing the "Estimate error tolerance" and "Determine map scale" exercises on the supplemental DVD.

Data design capabilities

As you consider the appropriate data design for your desired information products, it's important to be knowledgeable about the types of data that are available today, and the latest technology opportunities for data collection, presentation, and analysis. The following sections describe just a few data capabilities you should be aware of.

Survey capabilities

The GIS database can accept measurements from survey instruments of all kinds in three dimensions and execute all the traditional survey computations necessary to adjust those measurements and create coordinate points with known levels of error. Least-squares adjustment fit of disparate data can be conducted to get the best value for a point. Coordinate geometry (COGO) measurements can be included with the survey measurements and treated in the same manner. You can also add coordinates derived from GPS stations to the survey measurement database. These survey measurements and computations are carried out within the geodatabase in the same coordinate space as other vector and raster data.

The surveying data flow from field to fabric is greatly improved. Now one optimized operation can handle the process of moving from fieldwork station measurements through data processing of survey computations, to COGO and computer-aided design (CAD) systems for drafting and design, into GIS systems for integration with other data—all within one geodatabase in one coordinate space.

You can now integrate the survey measurements with the location of GIS features on the map, making the link between coordinates established from survey measurements and points on features. Thereafter, the features can be moved to their correct positions and stored in the database that way. Snapping tolerances, configuration algorithms, and batch processing of adjustments can be selected. All this enables you not only to improve the accuracy of the existing GIS database but also to add whole new GIS features defined by survey measurements. Display of error ellipses can provide measures of error of the new feature locations. The tolerance to relative error and absolute error (expressed in IPDs) can be quantitatively compared to the accuracy of feature placement.

Topology

Topology, an arcane branch of algebra concerned with connectivity, has always benefited the GIS industry, never more so than now. Since the earliest days of GIS, topology has been used to identify errors in the vector fabric of the database. Specifically, it could identify when polygons were not closed or when lines overshot the junction, when there were breaks in networks, when names had been incorrectly linked to features, or when two names had been assigned to one feature or no name had been assigned to a feature that ought to have a name. Creating a topologically correct dataset was a significant step forward in establishing the accuracy of a GIS database. It was essential before the days of inexpensive computer monitors (cathode ray tubes), when digitizers were working blind and errors were frequent.

Topology is an excellent tool for establishing spatial integrity and for error identification and editing. It is useful not only in identifying and correcting errors but also in spatial analysis, performing in minutes tasks (like checking for connectivity of street segments) that used to take GIS technicians days of poring over maps.

Current versions of topology operate in three dimensions. In these multilayered topologies, coincident or intersecting features or parts of features in one layer can be intelligently linked topologically with those from another layer.

In working with a geodatabase, technicians establish rules to control the allowable spatial relationships of features within a feature class, in different feature classes, or between subtypes. For example, lot lines must not have dangles, buildings must be covered by owner parcels, and so on. A topology is itself a type of dataset in a geodatabase that manages a set of rules and other properties associated with a collection of simple feature classes. The feature classes that participate in a topology are kept in a feature dataset so that all feature classes have the same spatial reference. A topology has an associated cluster tolerance that can be specified by the data modeler to fit the precision of the data.

You can apply topology to a limited area, which in today's very large databases gives considerable flexibility. Consider the advantages of such focused topology in error identification. Creating a topological rule does not ensure that it will not be broken, but does ensure that the error will be identified.

Areas of the map that have been edited and changed but not checked for topological consistency according to the rules are called *dirty areas.* When the dirty areas are checked they are said to be *validated.* Note that any errors found are stored in the database. These errors can be fixed using editing tools in the GIS; they can be left in the database as errors; or they can be marked as exceptions to the rule. This means you don't have to validate an entire dataset before using the database.

The editing tools for fixing errors have broad applications. Editing can be done on two feature classes at once. Features can easily be merged and split. Features in one class can be constructed based on the geometry of selected features in another class.

In summary, topological editing tools enforce spatial integrity constraints across multiple feature classes. The arrangement whereby the rules are established in a separate dataset, rather than being embedded in the data, allows for more flexibility within an organization.

The applications of these capabilities are substantial. Imagine, for example, incorporating new construction that changes the location of an associated road. The topological differences between the road centerline and the school district can then be identified and resolved. Now imagine this on a much larger scale. A city in Canada, for example, wanted to integrate all the features mapped on a citywide CAD system into a GIS for the city, potentially a time-consuming and therefore expensive process. The ability to use multilayered topology greatly facilitated this process, raising efficiency and reducing costs considerably. Continually available in the geodatabase, topology ensures accurate data in all layers as features, layers, and relationships continue to be added to or amended in the GIS.

Temporal data

Geographic information systems have not had extensive functionality for handling temporal data in the past, at best providing multiple static overlays. This is changing with the development of "time-aware" software that can now store time information as a layer property. Using this feature you can store information about your dataset that changes over time, and use animation functionality to visualize changing data and identify patterns. Real-time or near-real-time tracking of data is possible, as is temporal data analysis. Maps that display temporal data patterns allow users to dynamically visualize and analyze critical changes to data such as changes in land use, ocean temperatures, or the size and location of glaciers. Real-time tracking may include emergency response systems, threat detection, fleet tracking, or satellite tracking systems. A wide range of symbology can be employed, changing as the status of the event changes (as a hurricane intensifies, for example). With each playback, a histogram is available showing the number of events occurring over time, and this can be used for analysis of the events themselves (e.g., frequency of enemy shelling) or to determine the playback time or repeated playback times for further analysis.

Geographic phenomena that change over time can also be represented with graphical elements; for example, a line that charts the path of a hurricane. Roughly, it works this way: a simple event—designated with the ID of the object, the time, the place, and, if necessary, the status of the object—places a dot in space at a time in a certain condition. Several simple events can be linked to form a track.

The software also handles complex events, each of which includes additional information about the nature of the object being tracked. A dynamic complex event could be tracking a plane of a certain model, flight number, number of passengers, fuel load, age of aircraft, and name of pilot. Complex stationary events can be envisaged resulting from the use of traffic sensors.

Currently, the software handles moving points or points over time, as well as lines and polygons, all in a similar manner. Examples of what lines might represent in a GIS handling temporal data would include military fronts or weather fronts. Polygons might stand for satellite footprints or oil slicks, or temperature zones or precipitation measurements.

Temporal events can be dynamically presented using a time slider tool which shows a layer at various points in time. All of the displays can be animated for use on a media player and have plug-and-play capability with your own animation engine. Multiple animation tracks can be displayed at once; for example, if you want to simultaneously display temporal data at different scales, or showing different attributes.

It is possible to show temporal events (single or multiple) in the form of a clock. The clock wizard creates a circular chart, or clock, of temporal data, which can be used to analyze patterns of the data that may be missed when viewing the data on a table or map. Temporal data that illustrates specific aspects of the analysis, such as the nocturnal patterns of animal movement, can be produced simply by modifying the clock bands, changing colors, and using classes and legend settings.

With tracking analysis software, users can track near-real-time locations, up to 300 per second, and post the results to a server database. Multiple sources can be tracked and sent to multiple clients, including browsers and GIS applications.

Cartography

While GIS can produce elegant and beautiful maps (see figure 7.8), the process was once cumbersome, particularly when multiple maps had to be created. Thankfully, this state of affairs has greatly improved.

Now multiple maps can be made from a single geodatabase with a consistency of content and feature selection. This allows frequent updates to be cost-effective, as they are done once rather than on numerous files. Automation is achievable in the higher qualities of map sophistication rather than only in the lower mass-produced cartography. This use of intelligent cartography rather than "brute-force" cartography is starting to be applied to large-scale mapping operations.

Symbology catalogs now offer an impressive array of choices, including 3D representation, which is becoming increasingly expected by users. Annotation allows precise text placement and supports behavior rules that exceed the capabilities of standard labels. You can anticipate more sophisticated map element placement as well. Take the conceptually straightforward task of legend placement, for example: the rule could be, "place a legend somewhere over water or white space on the page such that it is no closer than 1.2 cm from either a coastline,

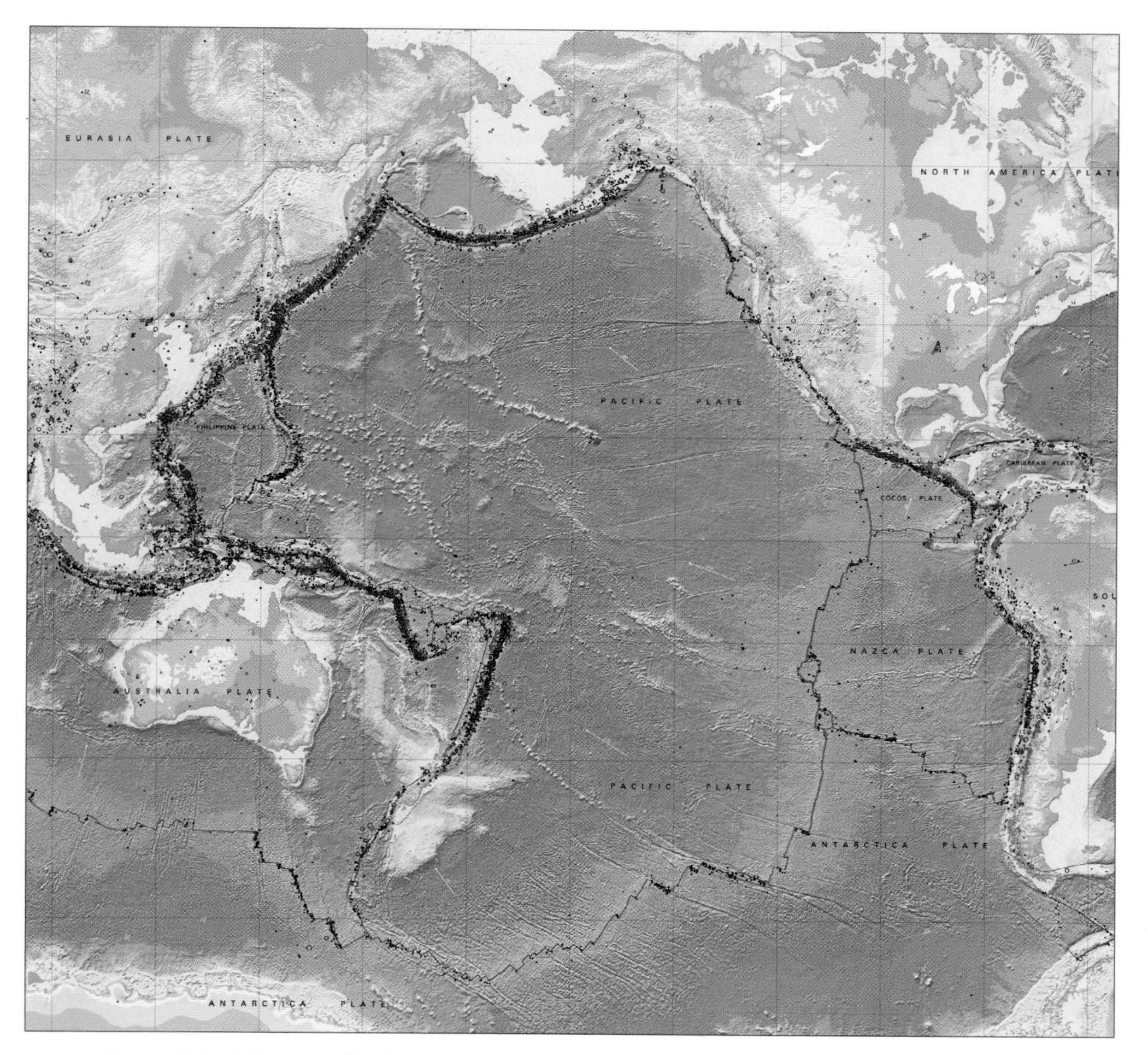

Figure 7.8 **GIS map of volcanoes, earthquakes, impact craters, and plate tectonics.**
Courtesy of the US Geological Survey, Smithsonian Institution, and US Naval Research Laboratory.

the edge of the page, or another element." Such a task is easy to define now.

Cartographic representation layers allow mapmakers to customize the appearance of individual features; for example, if you want a particular lake to have a different symbol than other lakes in the layer, or if you want to delete symbols of features without actually deleting the source data.

Automated page layout, text placement, generalization, and publishing continue to grow and to evolve. The most recent software features improved quality and functionality for labeling, with more options and greater flexibility for contour labeling, street number placement, callouts, curved graticules, stacked text, and geologic symbols. Automation of generalization is possible if models can be created to move from one known map design to a second known map design at a different scale. With the emergence of map templates and map book automation, the intelligent cartographic production of high-quality maps is becoming a reality for a growing array of map types in many industries.

Geoprocessing and spatial analysis

Geoprocessing—performing operations on GIS data to create new datasets that answer spatial questions—is simple with today's GIS software. Examples of geoprocessing operations include creating buffers around features, clipping one dataset so its extent matches another, smoothing and generalizing, creating topologies. An improved user experience includes a central, searchable toolbox with geoprocessng tools organized by function, the ability to create custom toolboxes for frequently used tools, geoprocessing environment settings (for example, desired output dataset location, processing extent, raster cell size, etc.) that can be applied to entire workflows or modified for individual tool outputs, and scripting options for advanced users. ModelBuilder automates the execution of multiple processes, with the output of one process serving as the input for the next process in the workflow.

Advanced geographic analysis is possible with software extensions. Spatial analysis allows you to model complex problems by applying calculations to multiple data layers. Variables can be weighted in the calculations, with an output of a rank or a probability (weighted overlay or fuzzy overlay). Advances in geographic science have been incorporated into the latest software builds. For example, new analytic tools allow multiscale spatial autocorrelation, space/time cluster analysis, areal interpolation, exploratory regression, and empirical Bayesian kriging. (If you don't know what these things are, don't worry, you probably don't need them.) For more specific types of problem solving, you can perform geostatistical, survey, network, tracking, 3D, and business analysis (network analysis is discussed below). A GIS is only as smart as the person or people behind it, so make it your business to understand the vast capabilities offered by GIS. You may start with an entry-level GIS program, but you can always adopt more sophisticated technology down the line.

Network analysis

The network dataset design has been improved, allowing for much more realistic modeling of network connectivity for the purposes of routing and tracing. This data structure fully supports multimodal networks, such as those found in transportation (car, bus, bike, rail, shipping, air routes) and hydrology (stream channels, roadways, wastewater, basins). Using a server-based architecture, network analysis and queries can be completed in real-time. For example, consider the real-time traffic reporting tool available on most smartphone mapping applications today.

The true value of the network structure is that it allows a flexible and extended query capability for the production of information products based on the network. With this query capability, commands take the form of instructions (set off by quotation marks, below) that can be almost as elaborate as you want to make them. The information products that are possible because of this extensive flexibility include the following:

- Shortest path. "Find the least-cost path (time, distance, etc.) through a series of stops." (Stops are network point locations.) Using online services, this is now available worldwide.
- Closest facility. "Given one location, find the closest (time, distance, etc.) facility (ATM, hospital, coffee shop, etc.)."
- Traveling salesman. "First find the optimal sequence to visit sets of stops, then run a shortest path."
- Allocate. "Build a shortest-path tree to define a 'service area' or a 'space-time defined space.'"
- Origin-destination matrix (OD matrix). "Build a matrix of cost between a set of origins (O) and destinations (D)." OD matrices are used extensively in many network analyses (such as Tour, Location/Allocation, etc.). Matrices are represented as a network.
- Vehicle routing. The core route optimizer. "Given a fleet of heterogeneous vehicles, in terms of capacity, available hours of service, cost to deploy, overtime costs, etc., and given a set of heterogeneous customers, both vehicles and customers with time window constraints, find the optimal vehicle/customer assignment and route for the vehicle."
- Location/Allocation. "Simultaneously locate facilities and assign (allocate) demand to the facilities."
- Optimal path. "Find the optimal path through a set of connected edges." (Garbage trucks, newspaper delivery.)
- Tracing. Works on a directed network, where "flows" in one direction, such as water, electrons,

FOCUS

Data structure for networks

Network connectivity can be based on geometry using a very rich connecting model, as well as database relationships such as air-flight relationships between airports or bus-route relationships to bus stops. These database relationships can be thought of as "virtual pathways" that allow for multimodal routing. This means that virtual people, commodities, or ideas can be transported by road, rail, ship, or airplane, transferring between each mode at known transfer points.

Turns are modeled as features, allowing them to be conflated between different feature classes. Turns can also be modeled as "implicit" turning movements, taking into account turning angles.

Networks are made of junction, edge, and turn elements derived from the geodatabase feature classes and object relationships. Elements have any number of attributes, such as travel time for cars, buses, emergency vehicles, heavy trucks, costs, restrictions, slope, number of lanes, pipe diameter, and so on. These attributes can be calculated from field values found in geodatabase tables or from a script. Attributes can be "dynamic," meaning they are calculated on demand. A dynamic attribute might query a recycling data structure containing current traffic conditions downloaded from the web.

Network datasets are versioned to support planning requirements and multiuser editing. Networks are incrementally built, meaning that only edited portions of the network are rebuilt rather than requiring a full rebuild on each edit.

wastewater, and so on, can be handled. Tracing tasks can be thought of as a general query: "Select what's upstream and downstream, find cycles, trace both up and downstream, or find dangles."

Imagery

Imagery, a form of remotely sensed data, is a type of raster data handled by a GIS. The raster data model is also used to store thematic data, features, and other quantitative information like elevation information. With a large volume of imagery, storage has been a serious challenge in the past (and still can be). It is not unusual for even modest collections of imagery to be 10s or 100s of terabytes in size, and large collections can easily run to petabytes and exabytes. The latest GIS software has integrated capabilities to manage, visualize, process and publish these large collections of imagery. Another aspect of imagery is that it is a source of derived data which results from feature extraction or geoprocessing workflows. This means that one image is often used multiple times, which can lead to duplicate storage if not well-managed. However, the modern GIS leaves the imagery in raw form when it stores it and processes only what is needed, on the fly, without storing intermediate data. This greatly reduces system storage requirements, particularly in a multiuser environment.

The GIS manages the imagery resource with the aid of another technical construct, the mosaic dataset, a catalog that points to and manages the raw imagery in place. The mosaic dataset is a database consisting of the metadata that is harvested from the original imagery. This easy availability of the metadata also allows the GIS to use the metadata to automatically enhance and position the imagery properly within the GIS for visualization. It also allows the user to make measurements directly on the images. These capabilities are available on workstations and desktops, but they also form the basis of publishing on servers that make imagery available in the cloud. For the user, this means that these large collections of imagery are easily navigated and they appear as one dataset that is intuitively accessed as one layer. Individual images are discovered by attribute or by time of collection. The on-the-fly processing provides custom processing and access to original pixel values for individual users accessing data even as an online resource from browsers or on mobile devices.

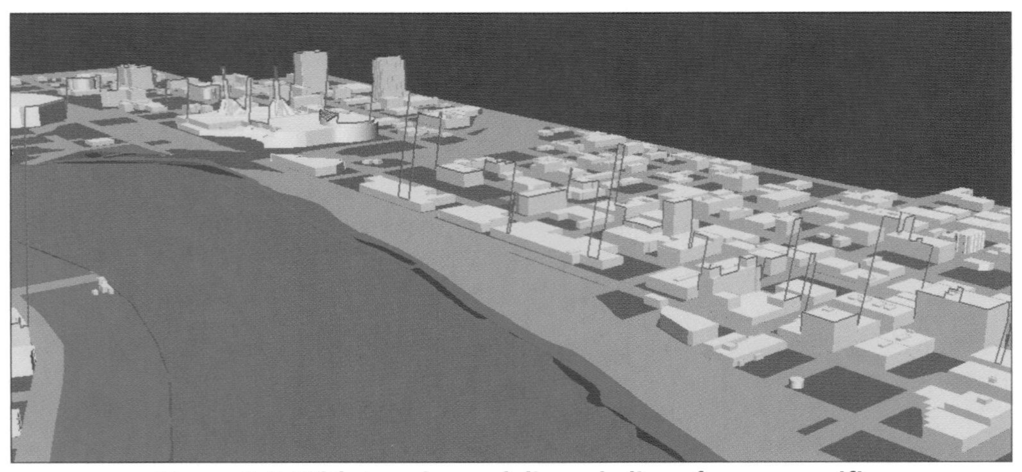

Figure 7.9 **With terrain modeling, skylines from a specific viewpoint can be generated and analyzed.**

Figure 7.10 **Many information products will have a mobile technology component for gathering, recording, editing, or serving data on mobile devices.** Maps courtesy of City of Redlands, San Diego Metropolitan Transit System and Transportation Management & Design, Inc., and Cobb County Water System.

Terrain modeling

Because massive amounts of elevation or other surface data can be stored in the database, terrains can be modeled with great precision in a GIS (see figure 7.9). National standards for lidar satellite measurements of height are being developed, which means very accurate representations of surface topography are becoming available. This opens the door to modeling other terrains, for example, showing the height dimensions of buildings in order to represent the skyline from any point on a 3D map. Spatial analysis can now take building height into account when performing line-of-sight calculations. Terrain datasets can be stored as triangular pyramids of TIN (triangulated irregular network) layers, with multiple resolutions for maps at different scales. Terrain measurements can be rendered on the fly when surfaces are accessed using a mosaic dataset.

Mobile technology

Many GIS systems rely on mobile GIS data acquisition and communication, and the latest, sophisticated mobile devices are sure to fuel the trend (figure 7.10). Interactive web maps and GIS functionality can be used by a range of mobile devices, including smartphones, Tablet PCs, GPS devices, and rugged handheld or keyboard devices especially suited for field data collection. As smartphones are becoming more ubiquitous, the number of users who view and search maps on their phones is steadily rising. Mobile GIS makes available an endless variety of information for different users, from the field analyst who needs to make sketches or take notes, to the ambulance driver who needs to be kept abreast of real-time traffic conditions and alternate routes, to a customer who needs to find the nearest storefront. Field data collection, mobile mapping, data editing, and real-time updates can streamline an organization's workflow, eliminate the need for paper surveys, and improve communication throughout an organization, as well as response times for public service departments. Examples of mobile applications include property damage assessment, location awareness and collaboration, city asset inspection and management, emergency response, and on-site data editing. If your information products require mobile technology, your GIS system should support a mobile workflow.

Improvements in mobile technologies impose more and more demands on GIS planning. Mobile devices are increasingly faster, smarter, smaller, and cheaper. They are also ubiquitous. Higher bandwidth and cheaper communication infrastructure is a necessary part of overall improvement. These recent achievements present two primary opportunities:

- First, it is now possible to interact simultaneously with many different devices through the cloud with its ability to support tailored APIs. This leads to improved communication and collaboration among large groups of people, in many circumstances, and perhaps with emergency concerns.
- Second, users can be effective data-gathering sensors, capable of providing a vast stream of fast and varied data.

These opportunities make the process of information product planning and design even more important. Product descriptions must now address design for different data scales and the smaller screens of mobile devices.

Community GIS

GIS is no longer limited to the desktop (or mobile devices) of a GIS specialist—now interactive maps are easily served to the public via the Internet, with simple user interfaces that make geographic visualization, navigation, and mapmaking available to novices as well as more experienced users. With the proliferation of cloud computing, organizations are no longer required to house their own expensive servers; data can be served from the cloud (off-site server farms), through the Internet. And users may not need to purchase expensive GIS software; the public can now browse enormous data catalogs, download data and apps, make maps, and explore geography in ways it couldn't even a few years ago. With free products like Google Earth, ArcGIS Explorer Desktop, and mapping apps for smartphones, GIS is available to the masses. If you think your organization will benefit from serving maps to the public, you will want to pay special attention to the chapter 9 section titled "Distributed GIS and web services." With the use of server GIS (either on-site or in the cloud), basemap caching, and adequate bandwidth and hardware platforms, you can share geographic information and provide location-based services to a wide audience.

An uncharted world of GIS collaboration is emerging. Using smartphones equipped with cameras and GPS, volunteered geographic information (VGI) has become popular, where any community member can accurately report incidents and problems such as potholes, oil spills, or species sightings. Organizations are partnering to form spatial data frameworks, a way to share spatial data in an accessible location (typically the cloud). Online mapping services like ArcGIS Online allow users to find data from all over the world, create and share maps, and communicate and collaborate with other members of the community. Community mapping programs provide a place for communities to contribute their regional basemaps to worldwide maps, including world topography, world street, and world imagery maps (see figure 7.11). These

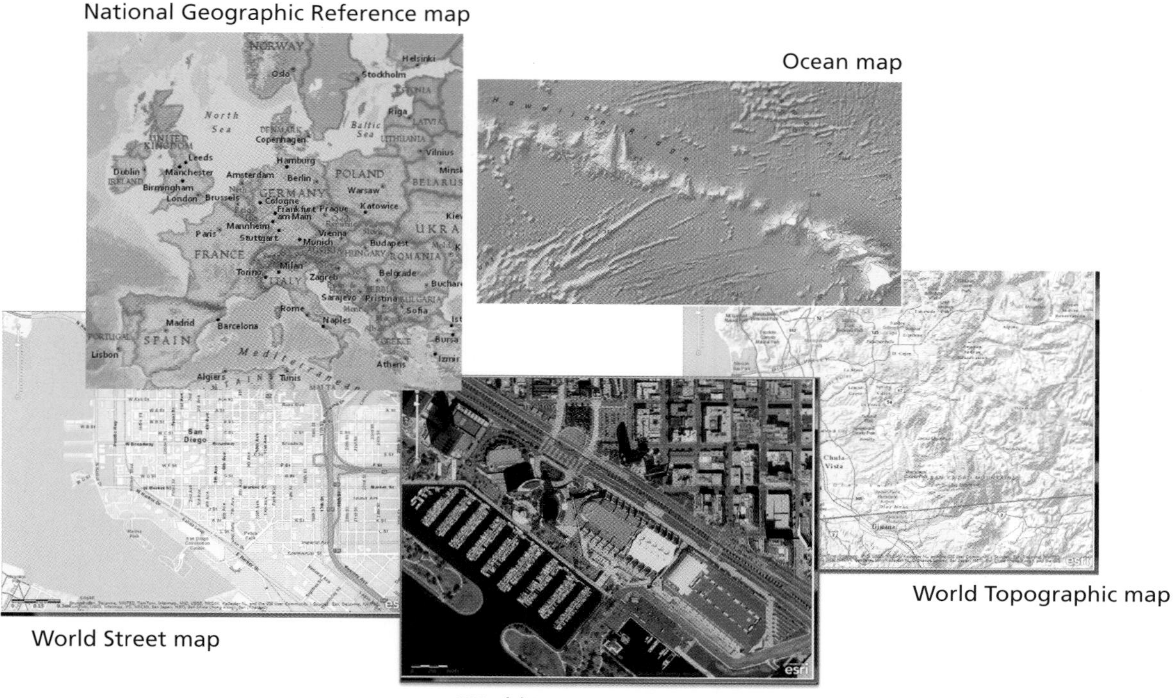

World Imagery International coverage at 1m resolution

Figure 7.11 **Free basemaps available from ArcGIS Online include world imagery, world street map, world topographic map, National Geographic world reference map, ocean surface map, and more.** Esri's ArcGIS Online.

maps are integrated into collaborative maps that can be accessed through GIS software or online, allowing users worldwide access to accurate, locally produced data.

Integrated services

Intelligent maps are the new medium. In other words, intelligent maps are the integration of data, maps, imagery, and models into applications where users can not only view maps, but *interact* with them. Today's maps provide information quickly; for example, when a user points to a particular feature, a pop-up window can appear showing relevant statistics or links to associated videos or photographs. Data can be edited and analyzed by multiple clients, and dynamic information can be seen in live time mode (a map displaying seismic activity as it is being recorded, for example). Server and cloud computing minimize storage limitations and allow huge datasets to be delivered to a large audience. Multiuser access, information sharing, collaboration, advanced mapmaking and analysis capabilities, and multifaceted geographic solutions are making today's GIS world an exciting place to be.

Who uses mobile GIS? A better question is, who doesn't? Public safety officials, utility workers, citizens, planners, executives, surveyors, field data collectors, market researchers, scientists in many fields...the list goes on. For many organizations, a mobile GIS workflow is essential.

Data logistics

This section highlights some important points to consider as you review your existing data and identify new data sources.

Digital data sources

From the early days of GIS until about fifteen to twenty years ago, most GIS databases were created with data converted from paper form, the slow and laborious process of digitizing paper maps. By now, more of what exists in the physical world is being measured and much of that data is now finding its way—in digital form—into the marketplace of information. Some of this data can be bought from commercial vendors for a fee, while other data is available for zero or nominal fees from public agencies.

Determining what datasets meet your requirements and sifting through them takes time—even knowing where to look and how to search can be a challenge. On the Internet, you will find a staggering array of map, tabular, and image data. Some of the best data is free, particularly from international, national, regional, and local governments.

Data portals, such as those available from the US Census (`http://www.census.gov`) or sites like the Geospatial One-Stop (`http://www.geodata.gov`), are emerging as good places to begin searching for reliable data. From private data vendors, prepackaged processed data is available for a wide range of applications. With the purchase of software, GIS vendors often provide worldwide basemaps and thematic maps that include vector and raster layers such as topography and elevation, satellite imagery, water bodies, political boundaries, landmarks, and streets for major cities. Maybe your own organization is a source of digital data. Access to this world is wonderful—in fact, oftentimes acquiring data is a cost-saver compared to creating data yourself—but no matter how easy data is to get, there is no shortcut around the imperative of recognizing its pedigree. There is no substitute for knowing how accurate and reliable your data is.

If you obtain digital data from other sources, reformatting and editing are frequently necessary to allow the data to be used with existing datasets. Most GIS software can work directly with a variety of data formats. Figure 7.12 shows the names and acronyms of some commonly used vector and raster data formats.

Standards for technology and data

Standards in technology usually refer to an agreed-upon set of guidelines for interoperability. Insofar as technology standards facilitate the effective sharing of application programs and data between offices, agencies, and the public, setting the standards to be met becomes one of the important requirements of a successful GIS. There are several standards to consider: operating system standards, user interface standards, networking standards, database query standards, graphic and mapping standards, and data standards. Here our focus is on standards regarding data, including digital data exchange formats.

Standards bring order to the seemingly chaotic development process and should be agreed on early in the GIS project. This applies to large enterprise-wide systems especially. Standards allow applications to be portable and more accessible across networks.

Along with the benefits, however, come the costs. Most smart managers realize that implementing standards costs time and money up front: to develop and implement standards, to train staff in meeting the standards, and to retrofit existing applications. As a conscientious manager, it is your duty to consider the true benefits and costs of your program and to communicate these to upper management so that funds will be forthcoming.

Another cost, aside from time and money, may come in the form of compromise in acceptable data quality and error tolerance. For example, if you establish a standard for positional accuracy that is plus or minus 40 feet, which is adequate for 95 percent of your applications, the usefulness for the other 5 percent of your

Vector formats	
ARC/INFO coverages	Etak MapBase file
Atlas GIS geofiles (AGF)	Initial Graphics Exchange Standard (IGES)
AutoCAD drawing files (DWG)	Interactive Graphic Design Software (IGDS)
AutoCAD drawing interchange file (DXF)	Land-use and land-cover data (GIRAS)
Automated Digitizing System (ADS)	Map Information Assembly Display (MIADS)
Digital Feature Analysis Data (DFAD)	MicroStation Design Files (DGN)
Digital Line Graph (DLG)	S-57
Dual Independent Map Encoding (DIME)	Spatial Data Transfer Standard (SDTS)
Esri file geodatabase (GDB)	Standard Linear Format (SLF)
Esri personal geodatabase (MDB)	TIGER/Line extract files
Esri shapefiles (SHP)	Vector Product Format (VPF)
Raster formats	
Arc Digitized Raster Graphics (ADRG)	ERDAS IMAGINE image format (IMG)
ARC/INFO GRID	Esri Grid
BIL, BIP, and BSQ	JFIF
BMP	JPG
DTED (Digital Terrain Elevation Data)	RLC (run-length compressed)
ERDAS	SID files
GIF	SunRaster files
GRASS (Geographical Resource Analysis Support System)	Tag Image File Format (TIFF)
Combined formats	
KML/KMZ (Keyhole markup language)	

Figure 7.12 **Commonly used data formats.**

applications will be compromised. You must develop data standards by taking into consideration the requirements associated with your information products, as well as the current standards and any anticipated standards that may be coming into effect in related departments or organizations.

The current, or established, data standards in your organization, if any, probably have been developed through informal arrangements, continuation of past practices, and the need to deliver information products. Most likely, they are undocumented and inconsistently applied. Existing standards are often out of date and fail to take advantage of the most recent technology.

Determining established and anticipated data standards is part of the design process. Again, much of the information you need will be in the IPDs. The GIS team, representing all participants, needs to arrive at a consensus on the following standards related to data:

- Data quality standards (i.e., the appropriate map scale, resolution, and projection for source material)
- Error standards (referential, topological, relative, and absolute)
- Naming standards (layers, attributes)
- Documentation standards (minimum amount of metadata required for each dataset)
- Digital interchange standards

In the United States, national data standards are being developed as part of the National Spatial Data Infrastructure. Take these into consideration as you develop your system. Some existing GIS standards are shown at the top of figure 7.13. After the GIS team completes its formulation of your set of standards, you

Geographic information systems standards	
Metadata	The Federal Geographic Data Committee (FGDC) content standard for digital geospatial metadata, ISO 19115
Data content	FGDC framework model, ArcGIS data models
Spatial data format	SDTS, GML, VPF, KML, shapefile, ISO-S57
Web-based spatial services	WMS, WFS, WCS, CS-W
Spatial data management	Simple features specification for SQL, OLE/COM, CORBA, ISO 19125:1,2 ISO 13249:3
Information technology standards	
Web services	XML, WSDL, UDDI, SOAP
Network protocols	NFS, TCP/IP, HTTP
Software API	JAVA, .Net, CORBA, COM, SQL
Emerging standards	Security: GeoXACML

Figure 7.13 **Standards relevant to interoperability.**

may find it useful, as some organizations do, to formally adopt and publish them.

In addition to meeting standards for data accuracy, you should make sure that the system you purchase adheres to the main IT standards of interoperability in general, shown on the bottom half of figure 7.13. The figure shows the main standards being frequently used today; you should know they exist and be familiar with some of the standard requirements.

Data conversion and interoperability

In order to get information from different sources into your GIS, you will use one or more data conversion processes that translate data from one format to another. A variety of methods are available. The one you choose will depend on the format and quality of existing data, the format of data from outside sources, and the standards you have set. Your basic options are to develop the data in-house, have the data prepared by an outside contractor, or reformat some existing digital data.

Usually in-house database development is accomplished by means of one or more of the following methods:

- Digitizing
- Scanning
- Keyboard input
- File input
- File transfer

You will likely include some of these terms in your list of functions to enter data, which is connected to your project's master input data list. Using data prepared by an outsource contractor is common in situations where the database must be developed rapidly, there is little or no in-house capability, or ongoing maintenance and further database development are not anticipated. When using an outside contractor for this crucial step (remember the "garbage in, garbage out" rule), take a close look at past experience with the vendor, its familiarity with the software and hardware you are using, and the availability of updates. Other GIS users with similar types of applications and database requirements are good sources of information about commercial vendors.

As the sources of digital spatial data proliferate, so does the practice of reformatting existing digital data into a format usable by your GIS. The latest software offers an extension that allows users to easily integrate multiple data formats into their GIS (data interoperability). It is

FOCUS

Finding data on the web

The Internet provides access to many public sources of digital data, some of which are listed below. On these sites you can search for downloadable digital data throughout the world or in the United States. Much of it is free.

- ArcGIS Online (http://www.arcgis.com) is a mapmaking service with access to various basemaps and local datasets throughout the world (subscription-based for multiple users).
- The GIS Data Depot (http://data.geocomm.com) is a data-sharing community with worldwide datasets available.
- The US Census Bureau (http://www.census.gov) provides downloadable demographic data for your area of interest.
- The US Geological Survey (http://www.usgs.gov) is a source for topographic maps for the entire United States.
- Geospatial One-Stop (http://www.geodata.gov) is a data portal for an array of US data.
- National Oceanographic Data Center (http://www.nodc.noaa.gov) is a source for bathymetry maps and more.
- The US Department of Agriculture Natural Resources Conservation Service Geospatial Data Gateway (http://datagateway.nrcs.usda.gov) is a searchable data portal for environmental and natural resources data.
- GeoGratis (http://geogratis.cgdi.gc.ca) is an online data portal hosted by National Resources Canada.

There are also many online companies that offer high-resolution aerial photography for a fee.

When determining whether or not to acquire digital data, you must be able to assess the history and quality of the dataset. If you do not know the data's content, source, age, resolution, and scale, consider it potentially unreliable for your purposes. You should expect to receive some metadata in the form of a data dictionary or data quality report from the provider or vendor. The metadata should provide pertinent background information about the data. Digital data can speed the process of developing your GIS, but only if you first understand what you're buying or downloading.

now possible to directly read at least one hundred spatial data formats, export more than seventy data formats, and even diagram and model your own spatial data format.

Interoperability is the term used to refer to this capability of systems or components to perform in multiple environments or to exchange data with other systems or components. In a server environment, interoperability means you can publish data once and it can be used by many clients, even if they use different GIS software. An increasingly important capability is "schema change" carried out during the transformation procedures. This allows algorithms for classification change to bring two datasets into the same schema and avoid the discontinuities at the join, particularly at administrative boundaries. This greatly facilitates large area/national/continental/global database building from local/regional inputs.

FOCUS

Search capabilities

Current GIS software has powerful built-in search capabilities for searching on local drives, on a server, or online. Desktop searches work by scanning indexed network folders and databases. This is another reason why maintaining metadata is important: the keywords, item description, and thumbnail elements of metadata facilitate the search function. If your organization has large amounts of data stored on a server, you will want to search there before looking elsewhere for the data you need. Server search, designed for searches within an organization's intranet, works similar to desktop search, except the indexing and search are centralized in a server. Online search (via ArcGIS Online) provides a hosted catalog of information to which users can upload their own resources or create their own groups, intended for organizations who want to share their data with the general public. Creating an ArcGIS Online group is also a way to share data within an organization or collaborate with other groups.

Nevertheless, each data conversion method takes time, brings an associated cost, and may degrade information content. The development of digital data exchange formats and the proliferation of conversion and translation software have made sharing data easier, but you may still encounter problems reformatting data from other systems. Don't assume that the data is easy to use because it is in a digital form.

At times you will find that the cost of reformatting data into something usable is not cost-effective. For example, polygon data from one type of software may be only graphic, with no associated attributes or topology (CAD data, for example). The cost of rendering this data into something suitable for your work may be greater than the cost of in-house digitizing from hard copy. (There is an interchange format to get data from CAD systems called DXF.)

Another potential difficulty is that you may not get adequate documentation for determining the accuracy, currency, source, or anything else about the data. This means you'd have no way of verifying whether the digitized data would meet your needs. Even if you are desperate for the data and believe the source to be credible, you should use such data only with extreme caution. Consider the following as a useful way to think of the integrity of data: 80 percent of data equals 20 percent of problems; 20 percent of data equals 80 percent of problems.

The free ArcGIS app is available for download at the Apple App Store, Google Play, Amazon Appstore, and Windows Marketplace.

Chapter 8

Choose a logical database model

The concept of object-oriented database models has brought a host of new GIS capabilities and should be considered for all new implementations. Yet the relational model is still widely used, and the savvy GIS manager will be conversant in both.

Ideally at this point in the GIS planning process, you understand the data elements needed to produce your information products so well that you have identified by name any logical links required between them. You know where your data is going to come from; you can picture its limitations as well as its dimensions.

You are ready for the next step—deciding how you're going to structure the data according to one of three logical database models: relational, object-oriented, or object-relational. You will be modeling the organization of the data after one or the other of them, whichever allows your data to be stored and manipulated most effectively to create your information products. Each model affords certain characteristics that would help or hinder you in setting up this management structure for your particular database. You'll compare what you know about your data (and what you need to do with it) against what you learn about these models to see which one is the best match.

The system's end users aren't directly concerned with this database modeling. But for you, the GIS leader, it becomes a significant concern at this juncture because the tools to manage your database—with which to add, store, delete, change, and retrieve your data—are software applications. Therefore, the model you choose for the organization of your database will factor into your decision on what software system to recommend to manage it.

The crux of the issue is that a logical database model must "describe" a complex version of the real world in a database. Such a model not only represents the data in computer logic but also describes the data in terms the computer can virtually "understand"—the model sets up its version of the real world with all its rules and orders that your data must follow. How much it will cost to build and use such a database will depend on how closely you want to model reality.

Determining the best logical model and then building it requires a decision, and now is the time to start thinking about it. Once you've examined the three types of logical database models, you can compare the advantages and disadvantages of each using the table in figure 8.10. Another table, figure 8.11, lists some basic situations alongside the capabilities characteristic of the model needed to handle them. Finding a situation that resembles yours, you may find it useful to consider the model(s) suggested for it. Just remember, though, there's no substitute for thoroughly understanding your options.

Essentially, data becomes useful when it's put together into the information you need to do your work. Putting it together becomes easier after it's been logically linked with the data associated with or related to it and then stored that way. Basically, each type of logical database model describes its own distinctive computer logic behind how to store this virtually connected data in the state of readiness required for GIS to create information from it.

The relational database model stores collections of tables that are associated with each other based on fields they share in common (e.g., tables that contain the same column storing values for a single attribute). In an object-oriented model, instead of as rows and tables, data is stored as objects or instances of a class (a group sharing the same set of attributes and behaviors). For an even newer model, the object-relational, GIS software enhances the structure of the relational model with some capabilities of the object-oriented model.

The relational database model

Currently, the vast majority of geospatial digital data in the world is stored using relational database models. In a database set up as a relational database model, the data is stored as collections of tables (called *relations*) logically associated with each other by shared attributes. The individual records are stored as rows in the tables, while the attributes are stored as columns. Each column can contain attribute data of only one kind: date, character string, numeric, and so forth. Usually, tables are normalized—calibrated to one another—to minimize

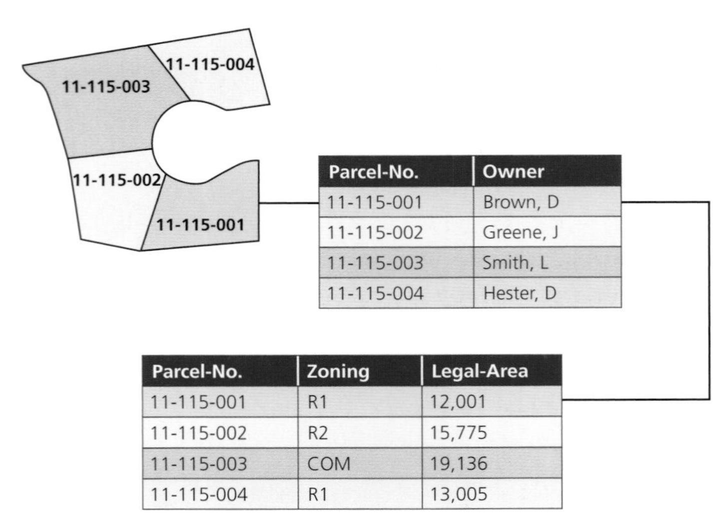

Parcel-No.	Owner
11-115-001	Brown, D
11-115-002	Greene, J
11-115-003	Smith, L
11-115-004	Hester, D

Parcel-No.	Zoning	Legal-Area
11-115-001	R1	12,001
11-115-002	R2	15,775
11-115-003	COM	19,136
11-115-004	R1	13,005

Figure 8.1 **Tables linked by a common field.**

redundancy. Storing spatial data in a relational database is greatly facilitated by using search engines such as SDE by Esri, which is software that allows the GIS to read spatial data stored in common relational database management systems (RDBMS), such as Oracle.

A GIS links spatial data to tabular data. In figure 8.1, you can see the links (shaded) between the spatial data displayed on the map (e.g., parcel numbers) and the table containing attributes about features on the map. A separate but related table uses the common Parcel-No. attribute column to link to the table showing owner information. There could be many additional tables available for use in GIS analysis and thematic display all linked back to one another by a thread of common fields.

Relational database models use a fixed set of built-in data types, such as numbers, dates, and character strings, to segregate different types of attribute data. This offers a tightly constrained but highly efficient method of describing the real world. Relational models execute and deliver results quickly, but require sophisticated application programming to model complex real-world situations in a meaningful way.

Consider the task of dispatching an emergency vehicle. Finding the quickest route requires detailed information about the direction and capacity of each street at different times of the day and the types of signals and change controls at each intersection. These complex variables require a large number of interacting tables, which the relational model makes possible. The strength of relational tables is that they simplify the real world and give swift and reliable answers to the queries that they can handle.

When you develop your logical database model using the relational approach, you may need to consider the layering and tiling structure of your data, also known as the *map library*. (There is a newer approach that uses seamless databases, but it doesn't work under the relational model. For that you need an object-oriented approach.) In this structure, a relational library of map partitions (or tiles) organizes geographic data into datasets of manageable size in an intuitively navigable tile system (see figures 8.2 and 8.3).

To set up the organization of data in accordance with the relational model you must consider all the factors involved in building the data layers of the map library: map units and projection, storage precision, attribute columns for the layered features, data sources and intended uses, logical linkages, data accuracy, and standards. Here is where planning pays off. Most of this information can be found in the MIDL (chapter 6).

Components of the relational model

The layering structure you choose for your map library is important; it will affect database maintenance, data query, and overall performance. You determine the content of each data layer as part of the conceptual database design. How you organize your data layers depends on how you

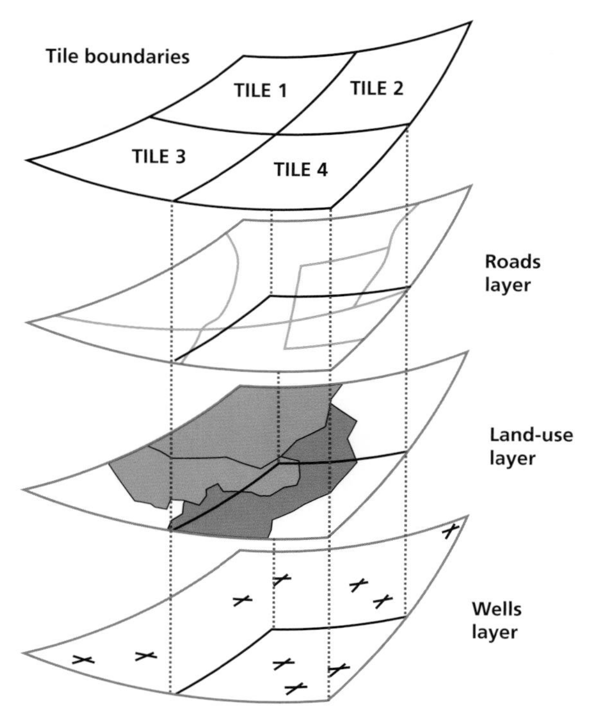

Figure 8.2 **Tiling: A way to organize datasets.**

will be using the data. You thought this through when you prepared the master input data list. (After reviewing the terms below, you can revisit your MIDL and make any necessary modifications.) In designing the layers, you will be taking into account many factors, some of which are defined here.

Layer

A layer is a logical grouping of geographic features (parcels, roads, wells, and so on) that can also be referred to as a *coverage* or *theme.*

First, review the MIDL to compile your final list of all the necessary layers. Early on, you should finalize the layer-naming convention. Settle on a unique (hopefully, descriptive) name for each and every layer. Once agreed on by the GIS team and published to the entire user group, these layer names should be adhered to rigorously throughout the rest of the conceptual database design. It's good practice to give layers descriptive names so that the contents can be easily recognized. The matrix in figure 8.4 shows a list of data layers from a typical municipal government GIS. It shows the layer name, the real-world objects represented by the name, the type of feature, and for each layer the attributes users need to see in their information product. You need to develop a matrix, probably much longer than this one, that describes all of the layers in your design. Keep in mind that it's not unusual for layer counts within an established GIS in a large municipal setting to number in the hundreds.

Tile structure

Review figure 8.2. The tile structure is the spatial index to the data in your GIS, and another thing that must be finalized before you can proceed with your design. Tiling speeds up access because the system indexes the data geographically, allowing you to search just the area of immediate interest and not the whole thing. Remember, GIS datasets tend to be large, so anything we can do to cut to the chase pays dividends. Once established, tile structure is labor-intensive to change, so think carefully about your tile selection. For example, it is better to frame your tiles using physical objects such as roads rather than on political boundaries that may change over time.

An abstract grid, such as US Geological Survey (USGS) quadrangle boundaries, offers a stable, somewhat standard, tile structure, too. If your IPDs show that your most important applications will need access by quadrangle, then making the same quadrangles your tiling unit could be beneficial.

Map projection

Use a single coordinate system for all the data layers in a map library. Data can originate from different projections but should be projected into a common coordinate

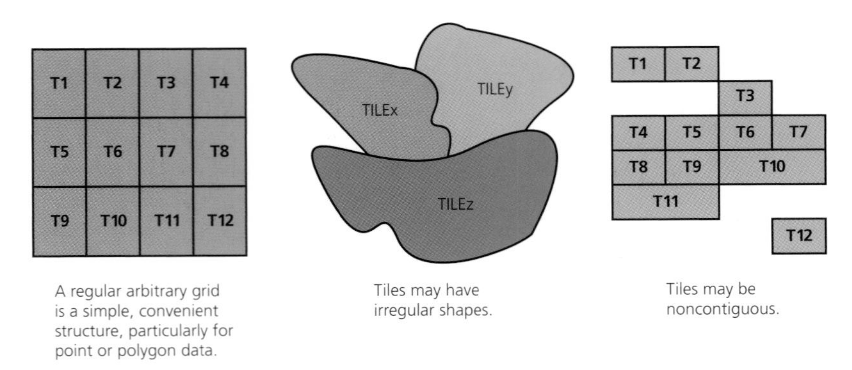

Figure 8.3 **Tile structure options.**

system before being added to the library. This is imperative because once you're in a mapping display mode, data with different coordinate systems or projections will not even appear in the same space.

Units (imperial or metric)

Use a single set of map units for your library. Sometimes the map projection and coordinate system you choose determines the units you employ. For example, the state plane coordinate system typically stores data in feet, while universal transverse Mercator (UTM) stores data in meters.

Precision (single or double)

The storage precision of x,y coordinate data is important. Review your IPDs to determine what precision is necessary for your organization. Coordinates are either single-precision real numbers (six to seven significant digits) or double-precision (thirteen to fourteen significant digits). The precision you choose also affects your data storage requirements because, as you might expect, double-precision requires more data storage capacity.

Features (entities)

If you can, organize features on the layers so that points, lines, and polygons are stored in separate layers. For example, you might store parcels (represented by polygons) in one layer, roads (represented by lines) in another layer, and hydrants (stored as points) in yet another. (You might also have cell-based grids representing images.)

Features should be organized thematically as well. For example, roads and streams could both be represented as lines, but it wouldn't make sense to represent them on the same line layer because they are different things and therefore need to be considered independently.

Attributes

Identify a set of attributes that pertain to the features in each layer. Say, for example, for each well site (a feature represented by a point), you want to know the following attributes: the well's identification number, depth, pipe diameter, type of pump, and gallons per minute. These comprise the data elements that need to be present and available to become part of your information product about this feature, so you will set them up as attribute columns in the layers.

Intended uses

You also need to know the intended use of the data. For example, if your city's water source comes from a series of public water wells, you should determine if you want only public wells in this layer or if you also want to include private wells. It is important to review your information product descriptions to determine the data requirements for each layer. If, in looking over your IPD, you are reminded that you need information about the public wells only, then you would not include data about private wells as attributes in the layers.

Layer name	Real-world object	Feature type	Attribute users need to see
Street	City streets	Lines	Name, class of street
Block group	Census block groups	Polygons	Age group population counts, median household income
Zone	Land-use zoning	Polygons	Zoning type
Land use	Actual activity of land use	Polygons	Activity type
Railroads	Main train lines	Lines	Railroad name
Sewer complaints	Incident locations	Points	Address, date, and description of incident

Figure 8.4 **Descriptive documentation of layers.**

Logical linkages

When designing your layers, make sure that the logical linkages—between the data layers and any attribute files necessary to make your information products—have been established (figure 8.5). It is important to look for linkages that apply but might not be in place. If you identify logical linkages not currently in the data but required for a product, you have to perform the necessary tasks to set up those links. For example, to link tables you can use a field they share in common. This common field, say "Parcel-No.," is referred to as a *key* when it is contained in both tables and can be used to link them together.

Source

You must know—and document in the metadata—the source of each data layer. The source information will affect the data standards set for each map library.

The object-oriented database model

Newer than their relational counterparts, object-oriented database models allow for rich and complex descriptions of the real world in a data structure that users find easy to understand. Objects can be modeled after real-world entities (like sewers, fires, forests, building owners) and can be given behavior that mimics or models some relevant aspect of their behavior in the real world. One simplistic way of thinking about the difference between this and the relational model is that the object stores the information about itself (all its attributes) within itself instead of in a bunch of related tables.

An object model of a street network, for instance, uses lines depicting streets. Each street segment shows behavior that models its real-world actions, such as traffic flowing in a particular direction, with the number of cars per hour able to traverse it at any given hour of the day or night.

Components of the object-oriented database model

When modeling your data structure after the object-oriented approach, you use the concept of class as a way to group objects that share the same set of attributes and behaviors—a very different template than the layering in the relational model's map library. Determining the classes that you need to organize your data is an important

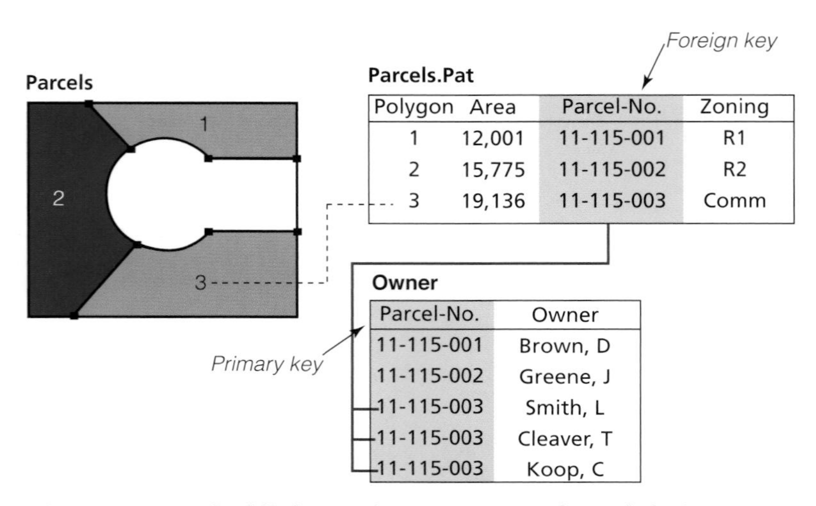

Figure 8.5 **Logical linkages between parcels and their owners.**

beginning step in this, a very different conceptual database design model, which offers a new way of thinking about objects. The components of the object-oriented database model are described below.

Objects

Objects represent real-world entities such as buildings, streams, or bank accounts in the virtual reality of the computer. Objects contain properties (attributes) that define their state and methods that define their behavior. Objects interact with one another by passing messages that invoke their behaviors.

Attributes

Attributes are the properties that define the state of an object, like the category of a street, the name of a building's owner, or the peak capacity of a storm drain.

Behaviors

Behaviors are the methods or the operations that an object can perform. For example, this virtual street may "know" how to calculate the increase in time needed to travel its length as rush hour approaches, or an account may know how to subtract money from its balance when a withdrawal is made. More simply, a river may only flow in one direction. These behaviors can also be used to send messages to other objects, communicate the state of an object by reporting current values, store new values, or perform calculations.

Encapsulation of behavior

By means of *encapsulation,* the essence of the object-oriented model, an object encapsulates (or encloses within itself) attributes and behaviors. The data within an object can be accessed only in accordance with the object's behaviors. In this way, encapsulation protects data from corruption by other objects, and also masks the internal details of objects from the rest of the system. Encapsulation also provides a degree of data independence, so that objects that send or receive messages do not need to be modified when interacting with an object whose behavior has changed. This allows changes in a program without the cost of a major restructuring, as would be the case with a relational structure.

Messages

Objects communicate with one another through messages. Messages are the act of one object invoking another object's behavior. A message is the name of an object followed by the name of a behavior the object knows how to carry out: "property/subdivision," for example. The object that initiates a message is called the *sender,* while the object that receives the message is called the *receiver.*

Classes

A class is a way to group objects that share the same set of attributes and behaviors into a template. Objects of a particular class are referred to as *instances* of that class. For example, the land parcel where you live is just one of many parcels that exist in your city, but each could be thought of as a unique instance. Unique as each one is, all these parcels share certain useful characteristics, like building type or zoning code; these shared characteristics are expressed as *classes.*

Determining the classes that you will need is an important step in your database design, allowing you to then diagram the relationships you need in your database model.

Objects stand for real things, but once they are part of a class they become more flexible than the real thing—like a bank account that can subtract money from itself. Objects are somewhat like capsules of data and behavior all rolled into one. Once you group each class into a template, the objects within it become instances of that class, retaining their uniqueness. Being part of a class, however, offers advantages of its own. Classes can be nested to any degree, and *inheritance*—gaining attributes and behaviors from another class—will automatically accumulate down through all the levels. The resulting tree-like structure is known as a *class hierarchy.*

Once you determine the classes, you can chart all the other ways objects can be associated with each other—in *relationships,* another important concept in the object-oriented model. Determining the classes that you need enables you to diagram the relationships you need; these *class diagrams,* in turn, allow you to diagram your database design.

Relationships

Relationships describe how objects are associated with each other. They define rules for creating, modifying, and removing objects. Several kinds of relationships can be used in the object-oriented data base model, including the following:

- Inheritance allows one class to inherit the attributes and behaviors of one or more other classes. The class that inherits attributes and behaviors is known as the *subclass.* The parent class is referred to as the *superclass.* In addition to the behaviors they inherit, subclasses may add or override inherited attributes and behaviors. A superclass is referred to as a *generalization* of its subclasses, and a subclass is a *specialization* of its superclass. For example, a house is a specialization of a building, and a building is a generalization of a house. A house class can inherit attributes and behaviors of the building class, such as number of floors, rooms, and construction type.
- An *association* is a general relationship between objects. Each association, in turn, can have a *multiplicity* associated with it, which defines the number of objects associated with another object. For example, an association might tell you that the object "owner" can own one or many houses. *Aggregation* and *composition* are specialized types of associations.
- Aggregation is a particular type of association. Objects can contain other objects, so aggregation is simply a collection of different object classes assembled into an aggregate class, which becomes a new object. These new composite objects are important because they can represent more complex structures than can simple objects, without losing the integrity of the simple object. For example, a building object can be aggregated with a sign object. But if you delete the sign, the building still remains.
- Composition is another specialized form of association. This is a stronger association relationship, in which the life of the contained object classes controls the life of the container object class. For example, a building is composed of (or contains) a foundation, walls, and a roof. If you delete what the building contains—its foundation, walls, and roof—you automatically delete the building, but not its sign.

Class diagrams

Class diagrams are used to diagram the conceptual database design according to the object-oriented model. They help you map the relationships that you need in a database model by illustrating the classes and relationships in your database. The following four class diagrams show you how to diagram classes, attributes, methods, and their relationships. The diagrams are a standard notation for expressing object models and are based on the Unified Modeling Language (UML), the emerging standard.

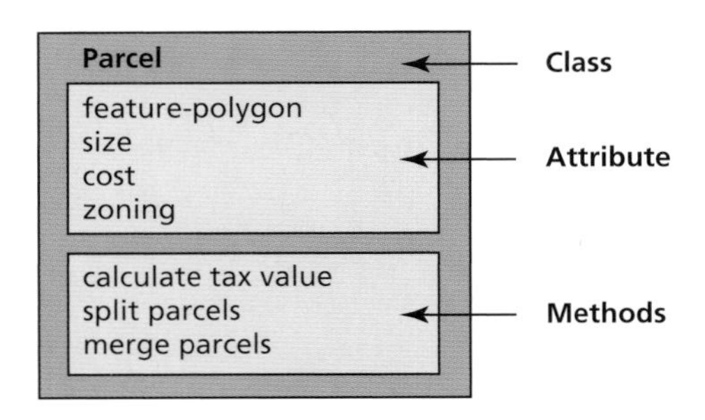

Figure 8.6 **What *class* can do.**

To review, a class is a set of similar objects. Each object in a class has the same set of attributes and behaviors (or put another way, the same set of defining properties and method or performance capabilities). In figure 8.6, the class named "Parcel" has attributes of feature-polygon, size, cost, and zoning, and it can perform these behaviors (or methods): calculate tax value, split parcels, and merge parcels.

Inheritance or generalization is the ability to share object properties and methods with a class or superclass. Inheritance creates a new object class by modifying an existing class.

Associations represent relationships between classes, as said earlier. The specialized types of associations, aggregation, and composition merit further discussion.

Aggregation is an asymmetric association in which an object from one class is considered to be a *whole* and objects from another class are considered to be *parts*. Object classes can be assembled to create an aggregate class. For example, the aggregate class "Property" can be created by aggregating the "Land parcel" class and the "Dwellings" class (figure 8.7).

Composition is a stronger form of aggregation in which objects from the whole class control the lifetime of objects from the subordinate class. If the object that makes up the whole is deleted, the subordinate objects that compose the whole are also deleted.

In figure 8.8 you have a "Network" class of water features with subordinate classes of "Stream" and "Canal." When you delete Network, the objects that compose it (Stream and Canal) are also deleted.

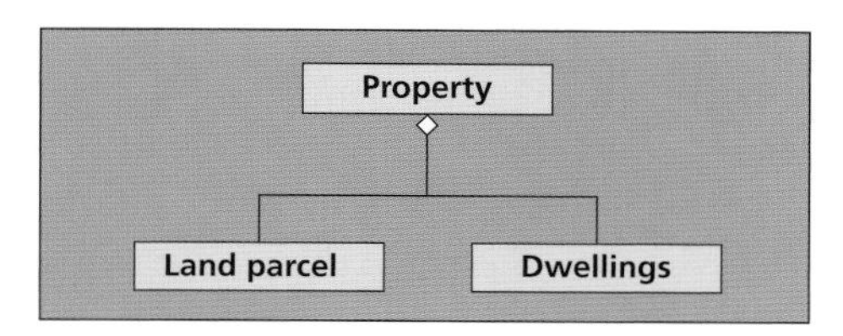

Figure 8.7 **Property as aggregate class.**

Multiplicity defines the number of objects that can be associated with another object. In figure 8.9 (which uses a notation device from the UML), we see that a water valve can have exactly one documentation reference and vice versa. In contrast, a pump station can have many pumps associated with it, but each pump can have only one pump station. While this might seem obvious, the rule system that this suggests is what gives object-oriented modeling its strength.

Consider the next line in figure 8.9: a pole can have zero or one transformer, but a transformer does not have the option of zero poles. Finally, we see that an owner can own one or many land parcels, and a land parcel can be owned by one or many owners.

The object-relational database model

The object-relational model is the most recent development in the domain of logical database models. In such a system, GIS software extends the relational database

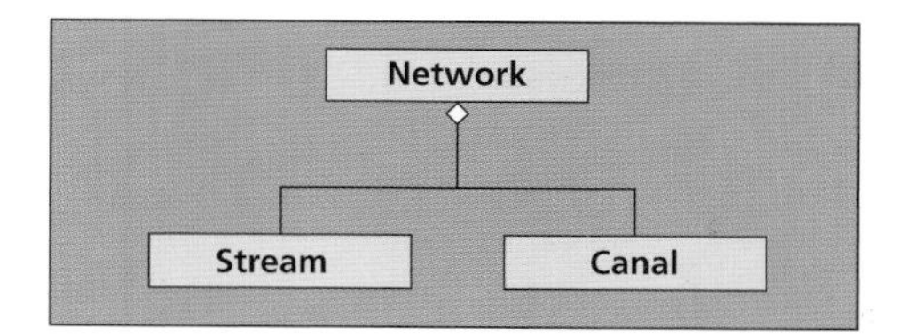

Figure 8.8 **Composition.**

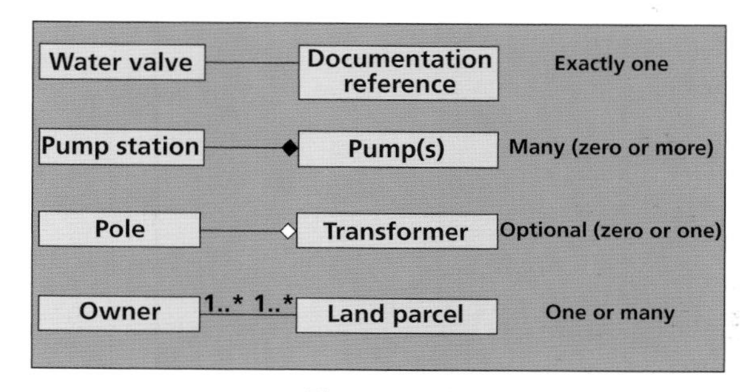

Figure 8.9
Multiplicity of associations defined.

to incorporate object-oriented behaviors that manage business logic and data integrity. This model blends well with other enterprise business systems because data is not encapsulated and remains in standard business tables supporting standard enterprise integration and management.

The object-relational model brings advantages of speed (important in large databases), the ability to handle complexity, and the database-building integrity of object-oriented designs. This model carries with it the additional advantage of supporting an extended form of Structured Query Language (SQL) and the ability to access typical relational database management systems. This alone can be an important consideration in enterprise-wide systems where other business applications need to access or contribute data to the GIS database.

Object-relational database models incorporate characteristics of both the relational and object-oriented databases. If you recall, the relational model uses tables with a fixed set of built-in data types (for example, numbers, date); in the object-oriented model, objects have unique attributes, and behaviors are encapsulated within the object. Enter the object-relational model. It extends the relational model by adding into some of the attribute columns a new and richer type of data structure, called the *abstract data type*. Abstract data types are created by combining the basic alphanumeric data types used in the relational model, or by storing a binary representation of the object in a BLOB (binary large object) field.

Abstract data types allow you to add specialized behavior to the relational model. This flexibility in data structure enables the object-relational database model to more closely describe the real world than can the purely relational model.

Object-relational software continues to improve and add more object functionality. The object-relational model is changing rapidly and taking more advantage of components of the object-oriented database model.

The geodatabase

The geodatabase (GDB), a storage container for geographic data, is an object-relational database model. Types of data that can be stored in the geodatabase include feature classes, attribute tables, or raster datasets. Many feature classes can be grouped into feature datasets. Advanced behavior, rules, and relationships can be

FOCUS

Ontologies

In information science, ontology is an organized description of the detailed properties of a feature or domain. An ontology can be thought of as an extension to a database model, for it illuminates the model by better describing complex relationships within and between datasets. There are formal ontological languages, such as OWL (Web Ontology Language), that assist in bringing structure to the task of arranging the descriptions. You can combine ontologies and dataset rules using the rule interchange format (RIF).

Ontologies

- provide extended metadata about a dataset;
- define the properties of a dataset;
- allow sophisticated queries and data mining;
- extend the ability to describe complex relationships; and
- may be published in conjunction with a dataset.

established to model sophisticated collections of data, such as network datasets, mosaic datasets, and topologies. It can be used by desktop, server, or mobile GIS applications.

Depending on the amount of data you need to store and the workflows of your organization, you will choose either a single-user geodatabase or a multiuser geodatabase (figure 8.10). For the single-user geodatabase, you will choose between the file geodatabase, with data storage capacities that can scale up to 1 terabyte (TB) per feature class, or the desktop geodatabase configuration using SQL Server Express, with up to 10 gigabytes (GBs) of storage and the added capabilities of geodatabase replication and versioning. The multiuser configuration is best for organizations and enterprises that will have multiple users accessing the geodatabase at the same time. Depending on the size of your organization you can choose either the file geodatabase (1 TB of storage, read-only publishing database), a work group geodatabase (10 GBs of storage, allowing 10 concurrent editors and readers, using SQL Server Express), or the enterprise geodatabase (unlimited storage, unlimited concurrent editors, choice of servers).

Advantages and disadvantages

All these models have advantages and disadvantages, as illustrated in the table comparing them (figure 8.11). How much these characteristics qualify as strengths or weaknesses will depend on your specific needs. In determining what database model is appropriate for your

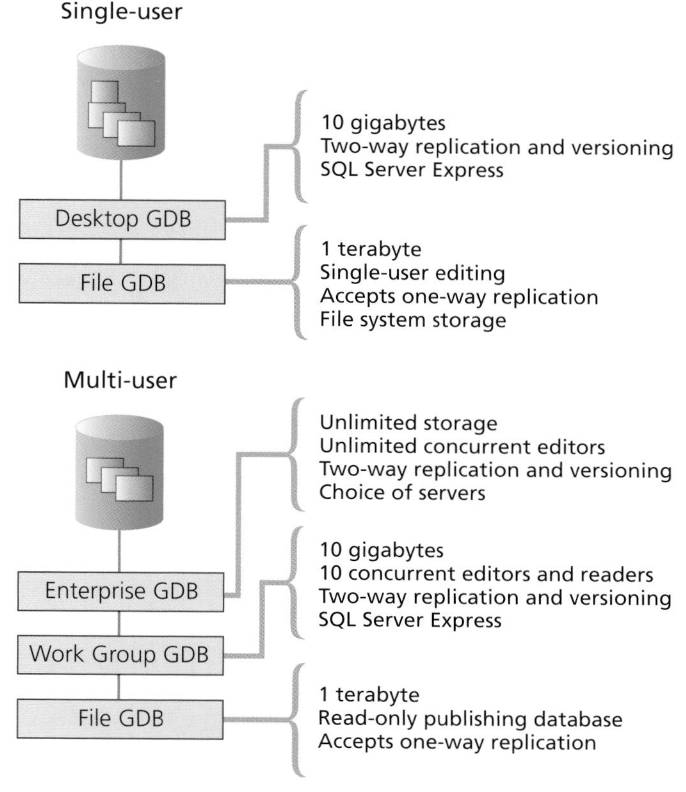

Figure 8.10 **Types of geodatabases and their defining characteristics.**

GIS, you must consider the types of functions required to create your information products. For example, if you need to perform a lot of network analysis with complex intersections or junctions, you could choose an object-relational model. If your information products do not require the type of functionality provided by object-oriented capabilities, you might choose the more established relational database model.

Choosing which logical database model is appropriate for your organization is not always a straightforward task. Much has been said about the object-oriented and object-relational models and how they will continue to affect the traditional relational implementation of GIS. However, this decision should be based on more than the conceptual differences.

Most GIS databases were initially based on the relational database model. But given the difficulty faced by this model in describing the complex behaviors of real-world objects, some GIS software vendors have opted for the object-oriented and object-relational models to support these more complex data structures. Choosing which model is most appropriate for your organization will in part determine which software you purchase. If a primary need is to model complex data relationships, then you better have a system that supports it. It boils down to the question of which type of database is most suitable for the work you want to do. Figure 8.12 describes some typical database modeling situations, along with the logical model characteristics needed to handle them, and suggests models that would be appropriate.

You should always consider costs before deciding what to recommend to upper management, and relative costs may well factor into your assessment of which logical database model will best meet your needs. Comparing costs of your high-priority information products is one standard approach. Set up two categories of costs for each database model, and then estimate the costs for developing your priority information products under

Database model	Advantages	Disadvantages
The relational	• The simple table structures are easy to read. • The user interface is intuitive and simple. • Many end-user tools (i.e., macros and scripts) are available. • New relationships, data, and records are easy to add and modify. • Tables describing geographic features with common attributes are easy to use. • Attribute tables can be linked to tables describing the topology necessary for a GIS. • Direct access to data provides fast and efficient performance. • Data is independent from the application. • Relational databases are optimized for GIS query and analysis. • Large amounts of GIS data are available in this format. • Many experienced developers, developer tools, textbooks, and consultants are available to guide users in working with this type of database model.	• Representation of the real world is limited. • Flexibility of queries and data management is limited. • Sequential access is slow. • Complex data relationships are difficult to model and often require specialized database application programmers. • Complex relationships must be expressed as procedures in every program that accesses the database. • A "performance penalty" exists due to the need to reassemble data structures every time the data is accessed. • Changes in a program may necessitate costly restructuring.

Figure 8.11 **Comparing three types of logical database models.** (continued on next page)

Database model	Advantages	Disadvantages
The object-oriented	• Users can create complex representations of the real world. • Users don't need to know the inner workings of an object because encapsulation, combining object attributes and behaviors, makes an object accessible through a well-defined set of methods and attributes. • Multiple levels of generalization, aggregation, and association are supported. • History in the database is maintained. • This model integrates well with simulation modeling techniques. • Multiple simultaneous updating (versioning) is possible. • Because it uses objects that occur naturally, the model is intuitive. • This model is well-suited for modeling complex data relationships. • This model requires less code in GIS programs, meaning fewer bugs and lower maintenance costs. • Data integrity is high because new data must follow the behavior rules. • Encapsulation allows for changes in a program without the cost of a major restructuring.	• Although object-oriented database models allow complex representations of the real world, complex models are more difficult to design and build. The choice of objects is crucial. • Import and exchange with other types of databases is difficult. • Some business applications may not be able to access or contribute to an object-oriented database. • Large and complex models can be slow to execute. • This model is dependent on a thorough description of real-world phenomena (particularly difficult in the natural world). • Object-oriented databases require the use of object-oriented computer languages for their analysis, but fewer people are trained in such programming.

Database model	Advantages	Disadvantages
The object-relational	• Fast execution. • Most features of object-oriented databases are available in object-relational. • Uniform repository of geographic data; allows use of legacy and non-GIS databases. • Data entry and editing are more accurate. • Data integrity is high (new data must follow the behavior rules). • Users can work with more intuitive data objects. • Simultaneous data editing (versioning) is possible. • History tracking and remote data replication can be accomplished. • Less need for programming applications to model complex relationships. • Makes possible the close integration with business model databases and the standard backup and support that accompany those models.	• This model is a compromise between object-oriented and relational database models. • Data encapsulation can be violated through direct SQL access to the data. • Support for object relationships is limited. • Complex relationships are more difficult to model than when using a pure object-oriented database model.

Figure 8.11 (continued) **Comparing three types of logical database models.**

each category for each database. Assign the various costs to either the database creation category or the application programming category. Under the former, include costs of developing and maintaining the database(s) needed to create the information products (e.g., costs of database design, data conversion, ongoing data exchange with legacy systems, staff training, and so on). Within the application category, list the costs of programming the application and any database manipulation required specifically for the information products. Use this approach with actual figures, if such data is available, or with relative rankings if not. What are the real costs of preparing twenty or thirty information products over the life of your system, using each logical database model for comparison? Once you've factored this in, you can make your choice of models and begin work on your conceptual database design. Next we explore the tasks involved in developing a

Data modeling situation	Logical model characteristics	Suggested logical data model
Forest inventory of tree stands, rivers, and roads for forest-harvesting analysis.	Simple relationships between features.	Relational
Addition of new forest stand attributes as forest matures. Addition of features such as property boundaries.	Easy modification and addition of new features and attributes as time passes.	Relational
Small staff. Need to minimize training and implementation time.	Simple, easy-to-use interface. Database that is simple and easy to design and build.	Relational
Enterprise-wide system must connect to existing sales and business partner databases.	Connects well to existing databases.	Relational or object-relational
Large business needs to perform suitable site analysis for locating new stores.	Need to use existing demographic data.	Relational or object-relational
Need to do real-time flood forecasting along a river system.	Complex representation of real world.	Object-oriented or object-relational
Need to simulate traffic flow through a street network during an emergency situation.	Integration with sophisticated simulation models.	Object-oriented
Major utility company needs to simultaneously update many parts of its large database as daily additions and repairs are made.	Multiple simultaneous updating (versioning).	Object-oriented or object-relational
Database is in constant use in mission-critical, life-dependent situations.	High level of data integrity.	Object-oriented or object-relational
Numerous new applications will be developed over time.	Low application development costs once initial model is developed.	Object-oriented
Complex analysis of natural resource features in large watershed (e.g., Columbia River Basin).	Fast to execute, particularly for large, complex analysis.	Object-relational
Many legacy relational databases and non-GIS databases to be linked to new GIS.	Links well with all types of databases.	Object-relational
Water utility needs to model water network, including water mains, laterals, valves, pump stations, and drains.	Complex relationships. Inheritance of attributes and behaviors.	Object-oriented or object-relational
Large amount of data maintenance and updating.	High level of data integrity.	Object-oriented or object-relational

Figure 8.12 **Suggested database models for some typical modeling situations.**

database, specifically under the relational and object-oriented models, and by inference, under the object-relational model insofar as it incorporates pertinent elements of both.

To review the concepts presented in this chapter, complete the following three exercises on the supplemental DVD:

- **Use a relational model to answer GIS questions**
- **Use an object-oriented model to answer GIS questions**
- **Use an object-relational model to answer GIS questions**

Chapter 9

Determine system requirements

Getting the system requirements right at the onset is what separates the professionals from the amateurs in successful GIS planning.

Now you will design a technology system—hardware, software, networking—that can handle the data and functionality you've specified. In chapter 8, you learned about creating the conceptual system design for your data; here you will create the conceptual system design for technology. Typically, these two design processes are carried out at roughly the same time because understanding the functionality you need helps you identify the technology required for it. In conceiving of a system design for technology, the focus is on defining a set of the technologies that will adequately support the demand for system functions in creating information products as needed. This is also the time and place to consider the implications of distributed GIS and web services.

This chapter's objective is to describe the relationships between the basic components of a technology system, in order to help you think about GIS within the context of your own organization's specific needs. Each GIS project is factually unique, from a hardware, software, data, networking, or personnel perspective, and so both the system design and the approach to configuring it must be tailored to fit. This chapter discusses a few options for planning a system technology, but every organization is different—there is no set recipe for integrating GIS into the workplace—so you must exercise your own professional judgment when presenting your recommendations to management. We will guide you along the way toward making your own choices.

First, we set the stage for workstation and software selection. Then, after examining some architectural options, a detailed example (a two-year "case study" of the fictional City of Rome) will illustrate an approach to identifying platform and network loads during peak operations. Finally, platform sizing and pricing models will show you a way to estimate the cost of the selected hardware, information you will need to include in your *preliminary design document*. Documenting your overall concept is an important step if you are to proceed with software and hardware procurement and for the GIS implementation that follows.

You can read more about the preliminary design document in the Focus box on pages 157–159.

Scoping hardware requirements

At this stage of the process, you need to consider your hardware requirements. The basic configuration for computer systems can be thought of as a three-tier option:

- Servers: multiuser UNIX or Microsoft Windows server/workstations
- High-end workstations: Intel workstation with fast processor speed (per core) and lots of RAM for running core GIS software
- Standard workstations: Intel workstations with medium processor speed and adequate RAM; for web access to GIS resources and thin desktop-client applications

You need to determine workstation needs (what kind, how many) according to the amount of data and the complexity of the processing required, as well as data storage and security needs.

Data-handling load

Refer to the IPDs and MIDL to calculate a quantifiable data-handling load that a software and hardware system will be called on to support. You need a system with an adequate data-handling capability to meet all its users' needs. The data-handling load is an estimate of how much work the system is expected to manage, an assessment that considers processing complexity, peak number of system users, and the volume of data the system is expected to store and process. You can use the data-handling load as a guideline for determining the computational power and storage dimensions of the system that you'll need to implement.

The amount of computational work required to produce an information product depends on processing complexity and data volume. At this point in predicting data-handling load, rough estimates of processing complexity and data volume will do to approximate the number of "boxes on desks" required, that is, the number of workstations or terminals necessary for individual users to interact with the GIS. (The sizing of server platforms is addressed later.)

Workstation processing complexity is either high or low. You can assess complexity according to the number and frequency of advanced functions used (e.g., topological overlay, network analysis, 3D analysis). (In the lexicon at the back of the book, functions considered to be complex are signified by an asterisk.) You should already have assigned a processing complexity category to each information product, when you created the IPD.

Data volume, either high or low, is the amount of data the system uses in producing one information product. Make your estimate from the number of different datasets needed to create the product, weighted (mentally) by the volume of data (number of items and size of area) that must be extracted from each dataset to make the information product. All you have to do is assign a high or low evaluation. For some information products, this will be an obvious selection; for others you'll have to use your judgment.

Defining workstation requirements

Based on the processing complexity of the information products (which was determined when the IPD was written), it is time to choose the type of workstation that will be required for each information product. Consider two workstation types:

1. High-end (2012 example): Windows 7 Intel Quad-Core Xeon Processor i7-2600; 3.4 GHz, 8 GB

memory; 1 GB graphics card; 27-inch flat-screen display; approximate cost in 2012 = $1,520.

2. Standard (2012 example): Windows 7 Intel Quad-Core Xeon Processor i3-2120; 3.3 GHz; 4 GB memory; 1 GB graphics card; 19-inch flat-panel display; approximate cost in 2012 = $910.

Figure 9.1 indicates the suitable workstation type (high-end or standard) for different combinations of processing complexity and data volume.

Based on data processing complexity and data volume assessments and your own experience (or with the advice of a consultant), you can now estimate how much workstation time each information product will need. Multiply the total number of workstation hours necessary to create the product by the number of times the product is needed per year. You can use this calculation to make a preliminary appraisal of how many workstations of each type you'll need in each department.

Occasionally one information product will use all the available hours on one workstation. More frequently, several information products may be produced from a single workstation or be made at different workstations in different parts of the department in different locations.

Terminal servers (also known as client-server systems) are usually employed where the data-handling load is high and multiple users need access to the same data. They are particularly appropriate for very large datasets.

Typically, every GIS user will have a dedicated workstation, the type of which is determined by the information products he or she is responsible for creating and maintaining.

Data hosting and user locations

The locations of machines on which data will be hosted, as well as where users will operate from, affect network communication requirements and should be considered in advance. For the purposes of estimating, let's assume that an information product will be generated in the user's department (although in the era of web services, this may not indeed be the case).

If you are a single user in a single department with a single computer using either kind of workstation and not hosting a web site, this part of the analysis is extremely simple. On the other hand, if you need to consider the location of numerous databases, along with users with different types of workstations and in different departments, the situation is more complex. We examine this more in the "Distributed GIS and web services" section of this chapter. For each department, whether in the headquarters building or at a remote site, you need to calculate the following:

- The total number of workstations using GIS, first for high-complexity processing, then for low-complexity processing (if both, use high)
- The number of users who will be using their workstations concurrently at the time of peak usage

If any of the departments are hosting a web page based on GIS data, you also need to factor in the anticipated number of visits or transactions per hour; either measure is helpful. Transactions are best but more difficult to estimate. You should use one or both in your analysis.

You can put together your own estimates in a table, as in figure 9.2, which will aid you in assessing your network communication requirements. The example here, estimating usage for three departments in headquarters and two

Processing complexity	Data volume to be handled	Type of workstation required
High	High	High-end
High	Low	High-end or standard
Low	High	Standard
Low	Low	Standard

Figure 9.1 **Assessing workstations for information products.**

		Processing complexity				Intranet/ Internet
		High		Low		Visits or transactions per hour
Location		Total users	Peak concurrent users	Total users	Peak concurrent users	
Headquarters	Planning	8	2	16	5	250
	Engineering	12	5	5	3	
	Operations	2	2	30	10	
	Totals	**22**	**9**	**51**	**18**	**250**
Remote sites	Exning	10	3	21	6	500
	Gazeley	8	2	10	5	400
	Totals	**18**	**5**	**31**	**11**	**900**
	Grand totals	**40**	**14**	**82**	**29**	**1,150**

Figure 9.2 **Estimating peak network usage.**

remote offices, shows that 40 people will be considered high-use clients, with 14 of them needing simultaneous high-use access. Also, counting users from all locations, 82 will require low-use access, 29 of them needing it concurrently during peak hours.

It is also useful at this stage to make some notes on the computer system with which the GIS manager will access the system. The GIS manager usually has special superuser privileges and specialized mission-critical, rapid-response applications, and should be equipped with a high-performance machine.

Data storage and security

The type and cost of the disk space you'll require depends on the volume of data your system must handle and on how much you need to protect your data. You should be keeping track of data volume amounts within the MIDL. Use the total data volume recorded in the MIDL to estimate the actual amount of disk space required by the information products. Then add 50 percent more to allow for indexing.

Data volume can be calculated as the number of gigabytes or terabytes of storage space required and can be classified as shown in figure 9.3.

Data volume	Storage space required
High	Over 1 petabyte
Medium	10–1000 terabytes
Low	Less than 10 terabytes

Figure 9.3 **Data volume categories.**

Your organization may require some level of data security measures to protect against accidental or deliberate loss and corruption of data. Most organizations also restrict access to sensitive data to prevent misuse. A network-based mass storage system can be secured against loss, for example, by including mirror sites as backups as well as tiered user-name and password protection. Standalone personal computers offer much more limited security features. You would classify computer security levels from high to low:

1. High security: For example, full mirror, RAID (redundant array of independent disks) level 1/0 (highest) protection (100 percent redundancy of data on mirror; no need to rebuild data in the case of single mirrored disk failure).
2. Medium security: for example, RAID level 5 protection (a much lower level of protection than RAID level 1/0). Two disk failures within a single multidisk data volume will have an effect, and it is more difficult to rebuild a database in the event of a failure.
3. Low security: Tape or compressed disk backup (you're only as good as your last regular backup).

The appropriate amount of storage and security and what it costs are shown in figure 9.4.

Preliminary software selection

In concert with other system components, software programs orchestrate the computer operations necessary to input the datasets and make the information products. By adding up these functions—and highlighting those most crucial to your workflows—you can determine the functionality you need from a software program. The results of this summary and classification will give you the function requirements, which you can provide to the software vendor after your proposal is approved.

Basically, your plan must envisage software that does all the things you need it to do, and performs very reliably the things you need done well and most often. So you must specify and quantify those functions first, rather than assume that every GIS program will have them. You can buy the latest software available, but you're wasting your money if it is missing a key bit of required functionality. Or what if it lacks speed and efficiency in the execution of some key function? That can be equally critical. For instance, when you know you'll often need crucial data from the public works department, you'll steer clear of a software program that does not easily import data from there. Summarizing the type and number of functions you need, and then classifying them according to workflow and frequency, will keep your planning on track as you consider software selection.

Summarizing the function requirements

To summarize your function requirements, simply write a list showing each function alongside the number of times a software program will be called on to perform it in a given year. In chapter 6, you identified the functions required to get data into the system and to generate each identified information product. You already listed these function requirements in the master input data list and in the individual information product descriptions. Your job at this point is easy: review the IPDs and summarize the number of times that each specific function will be used to produce the full set of information products in a year. Add to this summary the additional data-input functions identified in the MIDL (things like data import and conversion functions). The result is called the *total function utilization;* a summary not only naming the particular functions to be invoked but also quantifying how many times in the first year each function will be performed.

For planning purposes, you may want to forecast ahead, especially if you anticipate your organization's need for certain information products changing over time. Refer back to the calculations for a five-year span in chapter 6, and include every function that you foresee being invoked at any time during that period, including all of the datasets that will be input in the first five years (typically all of them), plus any other basic system capabilities.

Another option is to estimate each function's use on a year-by-year basis. Such summaries—for year one, year two, year three, and so on—can be useful, especially if you expect to require limited functionality in the early

Data storage options	Disk space	Security	Approximate cost (2012 prices)
Enterprise network mass storage	5 terabytes to 1+ petabytes	High security	$0.50–$1.50 per gigabyte
Work group server	300 gigabytes to 5 terabytes	Medium security	$0.10–$1.00 per gigabyte
Personal computer	150 gigabytes to 2 terabytes	Low security	$0.10–$0.50 per gigabyte

Figure 9.4 **The price of protection.**

years of a system and a higher level as your organization grows over time.

Classifying system functions

Classifying system functions is a way of highlighting those most crucial to your operation, so that you can select software with the capabilities that meet your specific needs. One glance at the function utilization summary tells you that the functions used most frequently are the ones that carry out basic system chores. You should have already created a list of functions required to make information products as well as those required to input the master data (described in chapter 6). Whatever your workhorse functions are, consider them essential to your system's operation. Typically, you'll find one or a small coterie of data manipulation and analysis functions that have an extremely high frequency of use. These functions are essential; your system relies heavily on them. These are *class 1* functions. The software you ultimately select must be able to perform them and perform them well. Any system that does not offer optimally suited and operationally efficient class 1 functions should be automatically disqualified from consideration.

Recognizing the least-used functions is just as straightforward. Take a look at them in figure 9.5, clustered along the bottom on the right side. These rarely used functions can be critical to individual operations nonetheless, so they deserve a place in your system. Classify those that need to be present, but not necessarily efficient, as *class 3* functions.

Those remaining (in the middle of the graph) provide important functionality and are heavily used: class 2 functions. During benchmark testing of systems prior to a major procurement, class 2 functions are always thoroughly tested. They must be in place and they must be efficient. Any functions required for basemap generation are best considered class 2 functions.

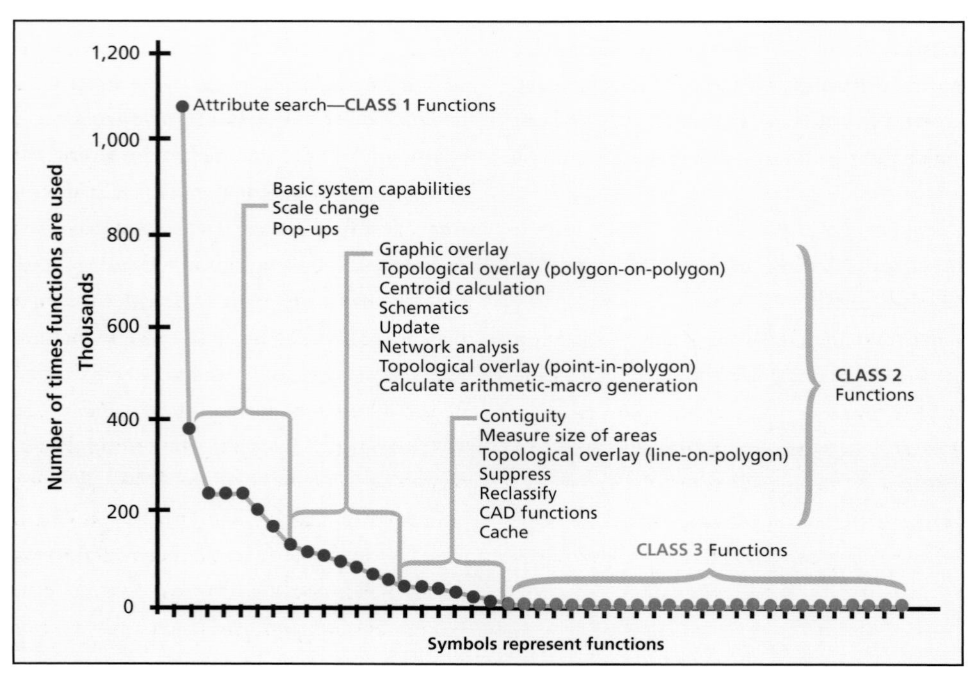

Figure 9.5 **Classifying functions that generate your information products.**

Read about benchmark testing in appendix B.

Document the classifications of functions required by your system overall, including the GIS and the frequency of their use. When you're ready to procure your system, you can provide vendors with this document—your list of objective criteria to support software selection. Note that only the functions on this list should be considered during procurement. This will help to keep you focused.

The figures you work with in planning are estimates, so don't be surprised when you discover they were off by 20 or 30 percent. You should expect variations in the methodology and information product timing over the five-year plan and accept this with equanimity. Recognize that some give is already built into the planning process—your technology design can accommodate a plus or minus 30 percent change in functional requirements. The objective is simply to avoid being 200 or 300 percent wrong and wasting thousands or even millions of dollars on systems that aren't even close, which is the track record of those who charge ahead without thinking about where they're going.

Interface and communication technologies

You already know the datasets by name and relationship, having traced their source in your MIDL; now it's time to get even better acquainted. Determining an adequate system interface and network communication configuration requires your understanding of where the major datasets in your and other organizations are maintained and stored, and how well they can travel to you and from you.

Choosing a system interface

You will probably need to link many of your organization's existing databases to your GIS. If you were to call on these datasets often, you would need each visit to be a quick trip and each conversation to be a free-flowing exchange. The determining factors for choosing a system interface are frequency and speed of access, and whether the data formats involved are compatible.

Take note of whether your GIS requires such frequent, repeated, high-speed access to major databases. Ideally, these would be two-way links that "talk to each other" easily: you're getting data and sending different data. Not every linkup will be a marriage made in heaven, though. Consider the city information systems folks who want to link their utility billing records to the GIS parcel database. If the two databases are stored in different formats, then to allow for this linkage you may need special interface software or a custom software interface, which requires programming. For your preliminary design document, you'll need to estimate in time and money how much the programming will cost.

You might consider using replication software or translation functions as an interface to the data sources. Establishing certain procedures to support frequency of data transfer is another possibility. Sometimes, all this discussion of database technology makes organizations realize they are using dated technology and triggers an organization-wide migration to a new standard database platform on which all applications, including the GIS, will successfully operate. In fact, GIS itself might be chosen as the integrating software. More and more it is.

Custom interfaces may be required less and less as service-oriented architecture (SOA) takes hold. The SOA strategy is for each business unit to provide web services; in other words, the interface between business systems or departments would be through standard web service protocols, rendering custom interfaces unnecessary.

Network communications

Network connections provide the communications link that allows you to both access data and distribute it throughout the organization. In a typical municipality, for example, a centralized server stores the data. City employees carry on with their workflows, getting the data they need from the server through their desktop computers. The network was probably configured to allow for these

FOCUS

Data capacity and data-transfer rates

GIS data is often characterized by large file sizes and shared through a network, which is simply a group of connected computers linked for the purpose of sharing. Largely, the purpose of a technology system for GIS is to move data quickly from here to there and have it arrive in the same shape it left in. Network bandwidth facilitates the former (data transfer from server to user), while the latter (processing) is left to CPUs or central processing units (often held on servers). Data is dependent on these components for quick travel.

In planning to implement a GIS, you need to consider your organization's requirement for both data capacity provided by the storage technology and data-transfer rates determined by the available bandwidth or network capacity for network traffic. Data capacity is how much data can be stored; at the moment, it is measured in gigabytes, terabytes, or petabytes (or even exabytes, zertabytes, and yottabytes). A server is capable of storing large amounts of data, but like your own desktop computer, the server can hold only as much data as its disk space permits. A small municipality may require only 400 to 800 gigabytes, for example, of disk space to meet its data storage requirements. Large data repositories, like the one maintained by the US Census Bureau, may have the capacity to store data in the petabytes.

Regardless of how much data can be stored, its size is useless if it cannot be put into motion and shared with many users in a reasonable amount of time. This is where data-transfer rates come into play, or how fast data traffic can be sent by the server over a network. Adequate network bandwidth capacity must be available to support these data-transfer loads at their peak.

Note that data traffic rates on the network are measured in bits per second. When translating the data volume stored on one storage disk to data traffic over a network, the formula is 1 megabyte equals 10 megabits (includes 2 bits of network protocol overhead for every 8 bits of data). Fifty-six Kb/sec is the same as 56,000 bits/sec; so 10 Mb/sec is the same as 10,000,000 bits/sec. Thus, a 10 Mb/sec local area network, or LAN, can support the transfer of nearly 180 times more data per second than a 56 Kb/sec wide area network or WAN.

This is often a misunderstood issue because of misinterpretation of communication units. But the devil is in these very details, especially as you approach planning for a big enough network bandwidth. Really understanding what the data-transfer rates mean is a critical dimension of GIS planning. Data-transfer speed is always the same (the speed of light); only the bandwidth, or throughput capacity, is different. Throughput capacity is very important; as cumulative network traffic approaches 50¬ to 75 percent of throughput capacity, significant communication delays will occur due to network contention at higher transmission rates.

existing workflows. But in doing its job of producing information products that streamline workflows, GIS will change how some things operate.

Whether you're thinking about a GIS in only one department or throughout the organization, implementing it is likely to have a major effect on the existing network. Be aware that GIS requires moving large amounts of data and complex (big) applications over a network that must allow for such heavy traffic. Providing shared communication segments that work like city roadways, networks can support only one communication packet at a time. So in order for a network to accommodate multiple communications during peak traffic loads, these data packets must wait in line. It is frustrating for end users to know that the data is on the server, but they must wait minutes for its retrieval and regeneration. With the newer technology, such traffic delays are experienced only when cumulative network traffic exceeds roughly 50 to 75 percent of the available network capacity. Beyond 75 percent, however, the network becomes a system performance bottleneck, and users suffer under lengthening wait times like urban drivers stuck in rush hour traffic.

Couple high-data loads with frequently used applications and you've got enough to choke undersized network connections or render them too slow. Advancements in technology have created GIS servers intended specifically to minimize data-transfer loads. Instead of simply sending all the data, these servers do work on the data and dispense the results only.

With such servers running the applications themselves, your GIS users may be able to access data through the existing network or after some slight updating. Or you may have to adopt an entirely new networking system. Predict the bandwidth suitable for your GIS using the approach provided later in this chapter. Make sure your calculations take into consideration the growth of your organization. One way to be vigilantly careful is to anticipate the highest possible numbers—in all your calculations—and then add 20 percent to be safe.

There are two basic network types: local area networks (LANs) and wide area networks (WANs). LANs support high-bandwidth communications over short distances. They provide high-speed access to data typically within a building or other localized environment, such

WAN	Low speed	Modem speeds	56K (53) Kbps
		Frame relay - time share	56 Kbps
	Medium speed	ISDN	128 Kbps
	High speed	T1	1.54 Mbps
		Fractional T1	Time share T1
		Digital service lines	2.2 Mbps
		Business synch	6.0 Mbps
		CATV	6–8 Mbps
		T2	6 Mbps
		T3	45 Mbps
	Wireless (least secure)	Line of site to tower	40 Kbps – 14 Mbps
LAN	Wireless	In-house (100-1000 ft)	10–300 Mbps
		Local bluetooth (<30 ft)	1–3 Mbps
	Networks	Fast ethernet	100 Mbps
		Fiber-optic (FDDI)	100 Mbps
		ATM	100–160 Mbps
		Gigabit (Fiber-optic)	1,000–10,000 Mbps

Figure 9.6 **Network speed options.**

as a campus. WANs support communications between remote locations. For example, a file server in a city office building shares data with its field offices through a WAN. Using a different protocol, WAN technology usually has a much lower bandwidth than the LAN environment and is more expensive due to the cost of constructing the communications infrastructure for data transmission over long distances. The Internet, in essence, is a global WAN. Figure 9.6 compares WAN and LAN network speeds.

Additional information on standard workflow network design planning factors is provided in appendix C.

Client-server architectures

Most commonly, transferring data over a network is accomplished by some form of client-server technology: the client requests data for an application and the server delivers data to the application. The transfer of information relies on a common language that permits this two-way street between the sender and the receiver, called a *communication protocol.* The type of client-server architecture chosen for a system will dictate which communication protocol is used.

The four basic types of client-server architectures are described below, along with the protocols associated with them. Most often you will find the "heavy" users who want to make their own information products within the first two architectures, which support data transfer between the data server and the application client. The last two architectures—transferring map display from application server to display client—usually don't need as much bandwidth as the first two because there is less data to transfer. (Fewer bits of data are needed to display an information product than to create it.)

Central file server with workstation clients

A central file server shares data with computer workstations across the network. The application software resides on the workstations, which perform both data query and map rendering functions. With its high demand for bandwidth, this type of configuration is best deployed over a LAN, which offers greater bandwidth over shorter distances (i.e., on campus, within a local organization). Common disk-mounting protocols associated with this network architecture are NFS (UNIX—network file services), and CIFS (Windows—common Internet file services). All of the disk-mounting protocols are standard TCP/IP (transmission control/Internet protocols) communications.

Central DBMS server with workstation clients

A central database management system (DBMS) server also shares data with computer workstations across the network (figure 9.7). The application software resides on the workstations, where map rendering is processed. The DBMS retrieves data from the server, but transfers only the data required to support the client display. Compared to the previous configuration, this type significantly reduces demands on the network. Yet, because it still requires the transfer of large quantities of data between the DBMS server and the client application, this configuration is also best deployed over a LAN. (Search engines compress the data before shifting it to the workstation client, which decompresses it.) The standard TCP/IP protocols apply.

Centralized application processing with terminal clients

In this configuration (figure 9.8), the client application software is stored and run on application servers (Windows Terminal Servers, WTS) located in the computer facility. The terminal clients remotely display and control applications executed on the terminal servers. These are the "dumb" terminals that first brought computing to the people. The only data transferred from the server to the terminal is the resulting display environment, which significantly reduces network bandwidth requirements. This type of architecture lends itself well to WANs. Common protocols associated with this network architecture are RDP (Windows—remote desktop protocol), ICA (Citrix—independent computing architecture), and X.11 (UNIX—open Windows display protocol).

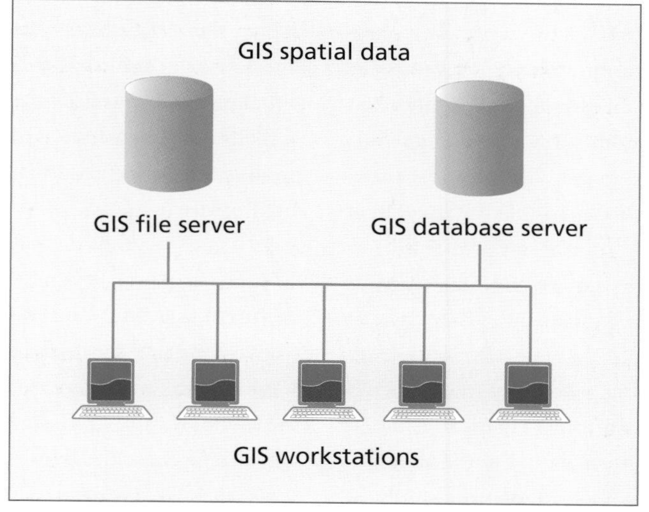

Figure 9.7 **Central server with workstation clients.**

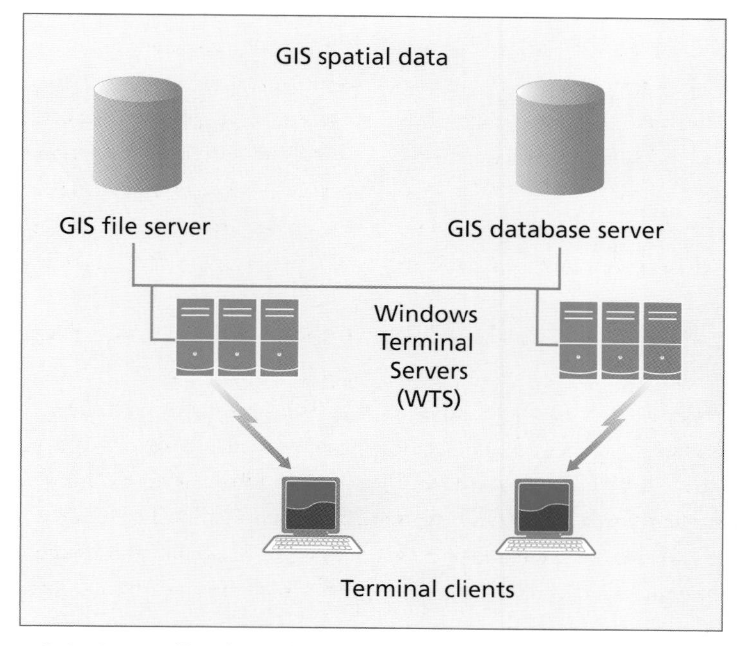

Figure 9.8 **Centralized application processing with terminal clients.**

Web transaction processing with browser or workstation clients

Web application software resides on servers (either web or map servers) located in the computer facility. The map server provides information products—data and maps—to web browsers or other "thin" clients (e.g., Java applications) via the Internet or a secure local intranet. This architecture (figure 9.9) allows a single application process to provide simultaneous support to a large number of random GIS user transactions. The protocol associated with this network architecture is HTTP (hypertext transfer protocol).

Every communication architecture has its own advantages and disadvantages. If the information products must be generated quickly through a WAN—for example, to create a locator map for emergency response—then a centralized application with terminal clients is viable. If application performance is not critical, but the ability to share the information with a large population of users is, then web transaction processing is appropriate. A central file or database server with workstation clients used to be necessary for users requiring high-performance access to large datasets that they must check out, edit, and then check back. But these days, terminal servers can provide almost the same high-performance computing as desktop. Keep in mind that many organizations use a combination of all four architectures to meet the demands of their user community. The application and data server options can be located in the cloud following the same architecture patterns.

General issues of network performance

Networking is a specialized and quickly evolving field; some issues related to network performance are outside the scope of this book. The following discussion about networking capacities is intended to give you an overview

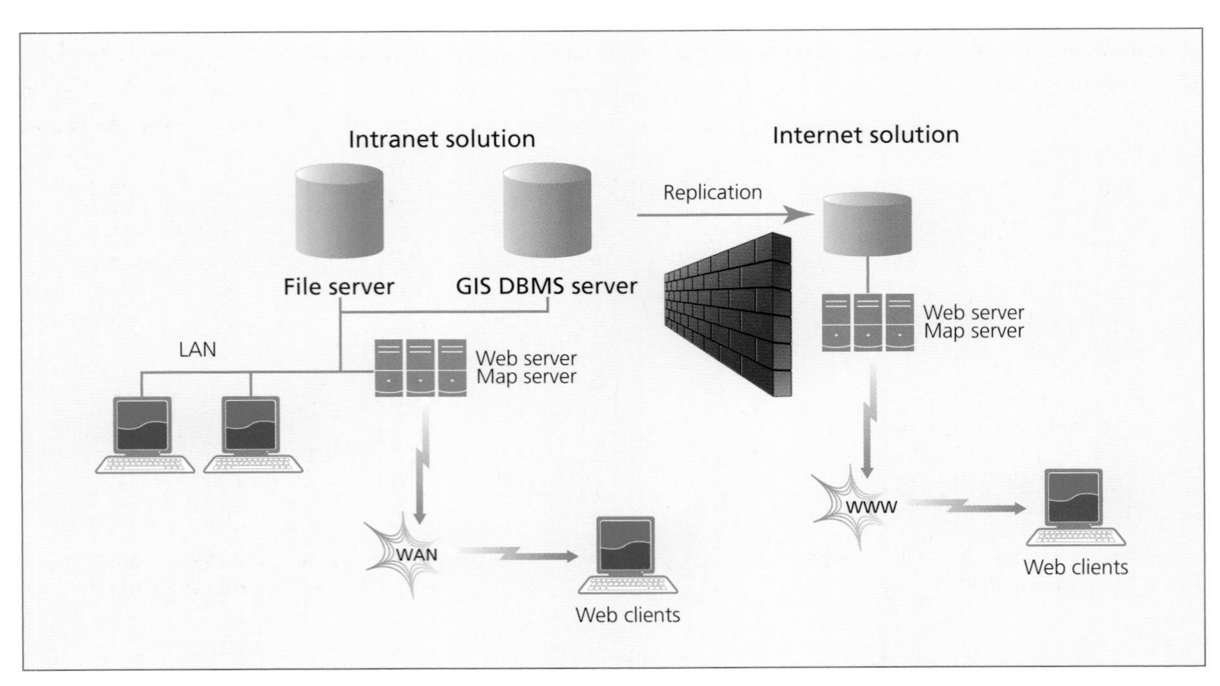

Figure 9.9 **Web transaction processing with browser or workstation clients.**

of network traffic considerations, not a working knowledge of the intricacies of network engineering.

You can use the volume of data transferred and network bandwidth capacity to identify the network transport contribution to expected user application wait times. A typical workstation GIS application might require up to 1 megabyte (MB) of spatial data that must be pulled over the network to generate a new map display or do an analysis. In contrast, a similar application deployed as a web service or to a terminal client might require just 100 kilobytes (KB) of data to support the same display or analysis.

Figure 9.10 compares the network time needed for the data transfer under various configurations. It measures the total transfer of data (network traffic) in megabits (Mb) that must be transmitted and factors in data compression percentages and traffic overhead that would typically be deployed under each client-server configuration. The transport time required to transfer this data is calculated for five standard bandwidth solutions. This chart illustrates how different client-server configurations affect the time it takes to transfer data.

The first two client-server communication options in the chart are examples of a central data server with workstation clients. In this case, the full 1 MB of data and traffic overhead is transferred from the server to the workstation to meet the demands of the application. The next two options represent web transaction processing, and the last, centralized application processing with terminal-client solutions. These latter three configurations make use of the data and application software on central servers located in the computer facility. While 1 MB of data is needed to generate the information product, the results are transferred from the server facility to the terminal clients as a display only, which reduces the amount of data required for transfer to 100 KB. Traffic is different for each protocol.

Of course, the third weighty factor to consider in configuring system architecture is the number of peak users or transaction rates. It is imperative to design a system that supports peak user workflows. Also, networks should always be configured with enough flexibility to provide special support to power users whose data-transfer needs exceed typical GIS-user bandwidth requirements. Figure 9.11 provides recommended design

	Client/server communications	Network traffic transport time (seconds)						
		Wide area network (WAN)				Local area network (LAN)		
		56 Kbps	1.54 Mbps	6 Mbps	45 Mbps	10 Mbps	100 Mbps	1 Gbps
Data	File server to GIS desktop client (NFS)	893	32.1	8.33	1.11	5.00	0.50	0.05
Data	DBMS server to GIS desktop client	89	3.2	0.8	0.1	0.5	0.05	0.005
Display	Web server to GIS desktop client (HTTP)							
Display	Light 200 KB => 2 Mb	36	1.2	0.34	0.04	0.20	0.02	0.002
Display	Medium 400 KB => 4 Mb	72	2.4	0.68	0.09	0.40	0.04	0.004
Display	Web server to browser client (HTTP)							
Display	Light 100 KB => 1 Mb	18	0.06	0.16	0.02	0.10	0.01	0.001
Display	Medium 200 KB => 2 Mb	36	1.2	0.34	0.04	0.20	0.02	0.002
Display	Windows Terminal Server to terminal client (ICA)							
Display	Vector 100 KB => 1 Mb >> 280 Kb	5	0.2	0.05	0.01	0.03	0.0028	0.0003
Display	Image 100 KB => 1 Mb	18	.06	0.17	0.02	0.10	0.010	0.0010

BEST PRACTICE

Figure 9.10 **Network time needed for data transfer within various configurations.**

guidelines for network environments based on the expected client loads. These guidelines establish a baseline for configuring distributed LAN and WAN environments.

Determining system interface and communication requirements

Now that you have a preliminary understanding of interface and communication technologies, you can think about the best configuration for your organization's system. Keeping in mind that you can find much of the information you will need for determining your system interface and communication requirements in the IPDs and MIDL, start asking the following questions about your proposed system:

- What external databases, if any, will be used within the system and what format are they in?
- What records need to be accessed? How frequently? How quickly?

If any external databases that you plan to connect to your GIS applications are stored in an incompatible format, you may need interface software in order to access them. The storage formats of all these datasets are identified in your MIDL. The records needed, the frequency of access, and the wait tolerance are given in the IPDs. Make a note of these databases and how often and how quickly your system will need to access them, so you can plan for obtaining the required interface software.

Next, ask these questions about your proposed system:

What are the wait tolerances of the information products?

Review the wait tolerances specified in the IPDs. Information products with low wait tolerances, such as emergency-service applications, will require a software technology solution that maximizes user productivity or one that takes advantage of interoperability technology.

Where is the data located?

Think about the location of the databases your organization will use with the GIS and review these data sources, listed in the MIDL. Ideally, all the data would be stored in one central database in a standard environment. In the real world, however, datasets are often stored on different servers throughout the organization. If this is the case and access to a dataset impedes the creation of your information products, you may find a solution in server

Local area networks bandwidth	Concurrent client loads			
	File server	DBMS servers	Windows terminals	Web products
10 Mbps LAN	2–4	10–20	350–700	150–300
16 Mbps LAN	3–6	16–32	550–1,100	250–500
100 Mbps LAN	20–40	100–200	3,500–7,000	1,500–3,000
1 Gbps LAN	200–400	1,000–2,000	35,000–70,000	15,000–30,000
Wide area networks bandwidth	**Concurrent client loads**			
	File server	DBMS servers	Windows terminals	Web products
56 Kbps modem	NR	NR	2–4	1–2
128 Kbps ISDN	NR	NR	5–10	2–4
256 Kbps DSL	NR	NR	10–20	5–10
512 Kbps	NR	NR	20–40	10–20
1.54 Mbps T-1	NR	1–2	50–100	25–50
2 Mbps E-1	NR	1–3	75–150	40–80
6.16 Mbps T-2	1–2	6–12	200–400	100–200
45 Mbps T-3	10–20	50–100	1,500–3,000	700–1,500
155 Mbps ATM	30–60	150–300	5,000–10,000	2,500–5,000

Figure 9.11 **Network design guidelines based on peak user workflows.**

technology (either on-site or in the cloud), which facilitates access not only to databases but also to software and ready-made applications.

Where are the data-handling locations (user sites), and what are the peak user workflows at these locations?

Data-handling information is crucial in determining how to tie users together through the network, and you already have it: you pinpointed the location of all sites expected to make use of the GIS data (figure 9.2). You also identified the number of users and the peak user workflows at these locations. You have gauged the traffic at rush hour—where it's coming from and where it's going. Now you need to assess if your organization has the communication infrastructure to handle all this.

What is the current network configuration?

Using your information product display requirements from the IPDs, along with the data-handling loads (peak user workflows), estimate the total volume of data (in megabits or Mb) that must be transmitted for your high-priority information products, paying particular attention to those with a low wait tolerance. Compare this demand with the existing bandwidth and traffic volumes for each segment of the network that will be used. Do you see a bottleneck in the making? Network access is vital to building and sharing information products—do not underestimate what your organization may need to accommodate the introduction of a GIS.

To complete the network configuration, you need to identify the location of each data source. Data can be accessed from a variety of data sources; some of your data may be provided by public web services or other organizations separate from your own local network. Connectivity between the client applications and the required data sources must be represented by your network architecture.

Your objective is to get a rough idea of what fits. Remember, you are merely seeking logical approximations intended to get your estimates within range. Your job is to identify the criteria required to make your system operate effectively, but you are not expected to be a systems administrator. So don't hesitate to consult with your network administrators, vendors, or others trained in this field and ask for their opinion. There are many different ways to configure a system. There might be more than one configuration option that could work well in your organization.

The remainder of this chapter focuses on a method for determining system requirements for distributed GIS solutions. If yours is a small project you may not find all of the information relevant. However, you may find the City of Rome case study, starting on page 135, to be a practical example about how to plan for the future, with any size project. Although the discussion specifies Esri software, it is applicable for most GIS products. For greater depth and fine-tuning, see *Building a GIS: System Architecture Design Strategies for Managers, second edition,* by Dave Peters (Esri Press, 2012).

Following the formulas for determining platform sizing and traffic bandwidth will give you the basic understanding of the process of sizing and traffic calculations. The Capacity Planning tool packaged with the *Building a GIS* book referenced above, will help you do the calculations. The tool can also be found on the System Design Strategies wiki site (http://www.esri.com/systemdesign). The following pages will help you understand how the tool arrives at the conclusions it does and will give you the necessary background to make adjustments to the tool's recommendations if you see fit.

Distributed GIS and web services

Distributed GIS, which allows more than one person to work on the same data in separate locations, is becoming an increasingly larger dimension of new GIS installations. Now that digital data is more available and the price of technology is decreasing, the distributed model is becoming more and more attractive. The main concepts, which the case study will illustrate, are peak usage and peak bandwidth loads, which enable you to assess bandwidth suitability, wait times, batch processing, platform sizing, and costs.

First review what you already know at this stage. For each department and business-process workflow, you have determined the degree of complexity of the information product and the number and size of datasets used. The same is true for the total number of users who will be making the information product, and, of those, the number that will be working concurrently. These have already been brought together in a list (figure 9.2) that, for each site in your organization, summarizes the total number of users and the number of high-complexity and low-complexity concurrent users, and the peak web usage expressed in number of requests per hour at each site.

To show how you might use such a list to model your own system design, we have built a realistic "case study" of a fictional city that we call Rome. This example illustrates one way (among several options) of estimating the platform sizing and the network bandwidth requirements of the system architecture configuration under consideration. (After walking you through the case study, we'll pick up again on several more factors to consider at this stage of planning on page 155.)

Platform sizing and bandwidth requirements

The calculations of the size of computers needed (i.e., the platform) and the availability of communications bandwidth required are based on estimates of the number of peak users on the system at one time. Each of these users will be generating processing and network traffic loads on system hardware and network components, based on the complexity of each user workflow and the user productivity. Standard workflow load profiles are established to support the design process. A workflow is a unique combination of system processing loads required to produce an information product.

User workflows

Platform sizing and bandwidth requirements will depend on identifying the GIS workflows that will be employed. This is facilitated by using standard workflows for common GIS software solutions. These standard workflows cover the most common GIS processes and can be customized to reflect variable display and database complexity.

Each standard workflow provides software component service times (processing load) that are combined into platform service times for platform sizing. Each standard workflow provides both platform processing and network traffic loads.

Workflow technology choices

The variety of user workflow choices continues to expand as software technology evolves. Technology today includes GIS desktop users, a variety of different web server clients, and a growing number of mobile clients.

Desktop workflows would be used for medium to high processing or display complexity. Normally this would apply to a smaller number of users; examples include GIS professionals, data maintenance, heavy business workflows, GIS project efforts, and data conversion.

Server (web service) workflows would apply to lower processing and display complexity and larger business and public user populations. Server workflows are also used for high-complexity repetitive tasks supporting a smaller group of GIS analysts—background geoprocessing services that can be deployed in a server environment. Server workflows are an optimum deployment pattern for published map services, geoprocessing services, and integrated business workflows.

Mobile workflows are appropriate for very low processing and display complexity, when mobility is required; such workflows often support many concurrent users. Mobile software provides loosely connected and disconnected operations, integrated and synchronized with enterprise operations.

Standard workflow baselines

Standard workflow baseline processing loads are predefined for the most common software technology patterns, derived from software vendor performance validation benchmark testing (figure 9.12).

Standard workflows provide baseline performance parameters for client display traffic (megabits per display, Mbpd) and client application processing time (client desktop, WTS Citrix application server, web application, GIS server [server object container, or SOC machine], search engine or mosaic dataset [SDE/MDS], and database management system). These software components represent the GIS software technology. Service times shown in figure 9.12 are relative to a selected 2012 baseline platform (Intel Xeon E5-2643 processor), where the published SPEC benchmark baseline per core is equal to 45.

SPEC (Standard Performance Evaluation Corporation) is a nonprofit organization with a mission to establish a standardized set of benchmarks by which to measure computer performance. The benchmark results referred to in this chapter are the latest available in 2012, published in the *SPECrate_int2006* benchmark (sometimes abbreviated to SRint2006). To find out more about the SPEC benchmark products, go to http://www.spec.org.

With workflows identified for each information product, peak network traffic loads can be determined by multiplying the client display traffic (Mbpd) by the number of concurrent users, times user productivity (average displays per minute per user). Dividing results by 60 seconds will give traffic in Mbps.

Similarly, the processing load on each software component in each workflow is combined into total platform

Workflow	Client Traffic	Software Component Service Times						Software Service Time
		Arc12 baseline		SRint06/core = 45.0				
		Design Model Metrics				Database		
		Client		Web	SOC	Data	Data	
Standard Workflows =========	**Mbpd**	**Client**	**Citrix**	**Web**	**SOC**	**SDE/MDS**	**DBMS**	**Total**
GIS Desktop =========	**Mbpd**	**Client**	**Citrix**	**Web**	**SOC**	**SDE/MDS**	**DBMS**	**Total**
GIS Desktop medium complexity	10.000	0.338				0.042	0.042	0.422
Remote GIS Desktop medium complexity	1.000	0.036	0.338			0.042	0.042	0.458
GIS Server =========	**Mbpd**	**Client**	**Citrix**	**Web**	**SOC**	**SDE/MDS**	**DBMS**	**Total**
GIS Web Mapping REST medium complexity	2.000	0.036		0.012	0.116	0.029	0.029	0.220
GIS Full Map Cache service	0.500	0.036		0.009	0.009			0.053
GIS Mobile services =========	**Mbpd**	**Client**	**Citrix**	**Web**	**SOC**	**SDE/MDS**	**DBMS**	**Total**
GIS Mobile client		0.166						0.166
GIS Mobile synchronization service	0.031	0.036		0.019	0.015	0.001	0.001	0.073

Figure 9.12
Examples of standard workflow performance baseline component processing loads.

usage, assuming a standard chip, peak concurrent users multiplied by user productivity provides the total throughput. Total throughput multiplied by component software service times gives baseline processing seconds for each installed server platform. Adding up all the workflow processing times gives the total baseline processing time for each server. Translating baseline processing time to selected platform processing time gives adjusted processing time for the selected platform. Dividing by 60 seconds gives the number of selected platform cores required for peak processing loads. Knowing the required platform core and core per platform gives the number of required server platforms and hence platform cost.

Custom workflows

System architecture design methodology has evolved significantly over the past several years. In 2009 we were introduced to standard workflows with estimated software service times for information product generation, and for use in platform sizing and bandwidth load calculations. An even more recent development (2010) provided the ability to refine standard workflows into custom workflows for each information product. This fine-tuning gives more precision and more accurate results, particularly for large systems. A complete discussion on developing custom workflow performance targets is provided in appendix D.

City of Rome case study

With the platform technology baselines (figure 9.12) and performance adjustment guidelines (see appendix D) in hand, put yourself in the shoes of the GIS planner for a large organization—a city—and follow along as this case study uses a simple method for determining platform sizing and network suitability for your system design plan. Later, with your own list in hand, you can use information you've already gathered to see if the method works for you in selecting a platform and in analyzing network bandwidth requirements based on projected user workflows. The results of your own analysis, perhaps like the one here, will provide a foundation for your hardware costs and documenting infrastructure implementation requirements. This will become a crucial part of the final report you will present to management.

During the business needs assessment, the city department managers identified specific end-user requirements (information products) needed to support their business operations. The first task is to collect these end-user requirements for both year 1 and year 2 and show peak user and web transaction loads projected for each business workflow. You will also need to identify the user locations and network connectivity to support your network suitability analysis.

Year 1 user locations and network connectivity

It is proposed that the GIS take advantage of the existing city communication structure for its first year (however, bandwidth capabilities will be evaluated and upgrade recommendations will be proposed as necessary to meet the year 1 objectives). All GIS users will be linked to the central GIS DBMS located in the IT department in city hall with search engine capabilities on the server so they can use the existing city system network configuration and bandwidth. Some remote users will link via Windows Terminal Servers and others through web services. Figure

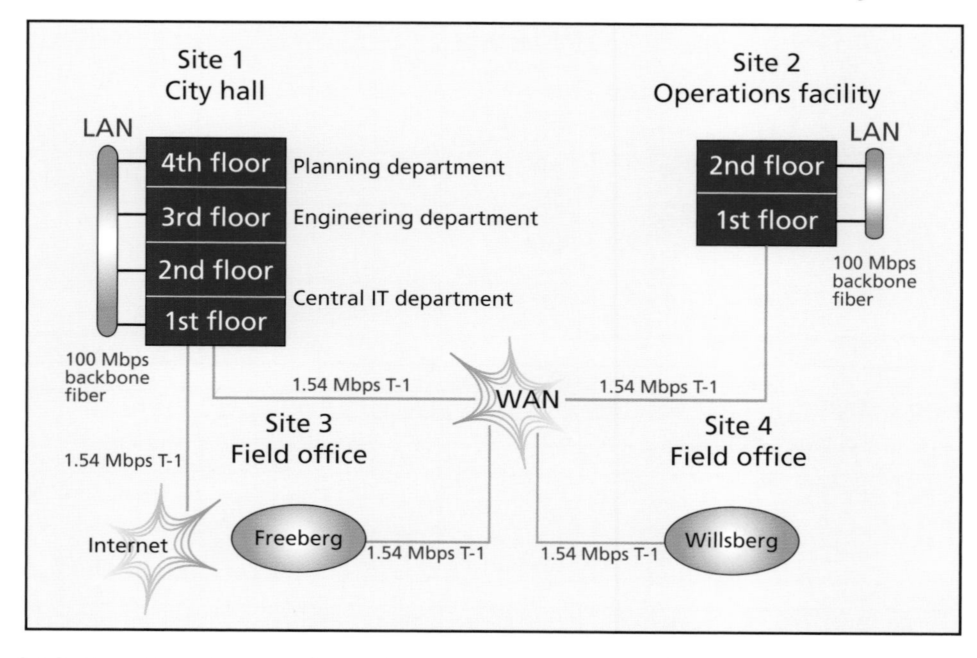

Figure 9.13 **Year 1 system configuration: user locations and existing network connectivity.**

9.13 lays out the user locations and network communications that exist at the start, the site configuration for year 1. This is the baseline reality from which the GIS will launch.

Year 1 estimated user requirements

The numbers indicating user requirements for the first year are listed in figure 9.14. Assuming the role of Rome's planner, you will be using these values later on, for comparison, in projecting how your system might grow as the city's needs change and grow. You will also use these numbers to estimate network traffic requirements during peak business operations. Tables like this one, in which you estimate the usage by department, are useful for planning purposes because you need to get department managers to sign off on the designs before going further in assessing the platform capacity and bandwidth needed. The user requirements table shows each department positioned within their site location (Planning, Engineering, and the IT department located in city hall). In the table, notice each information product has a unique number assigned to it for easy reference.

Year 2 user locations and network connectivity

Since you want to plan a system that is scalable—one that will grow smoothly as your users' needs grow—you must anticipate that growth so you can begin with network bandwidth and platform sizing suitable for expanding. So now you must estimate your organization's user

City of Rome - Year 1				Total Users	Peak Workflow Loads			
					Desktop (users)		Server (req/hr)	
Department	Workflow	IPD	User type		DeskEdit	DeskView	LocalMap	PublicMap
Site 1 - City Hall (LAN)								
Planning	Zoning	1.0	Planner	20		8		
		1.1	Web services				2,600	
	Permits	1.2	Inspector	20		10		
		1.3	Appraiser	15	8			
		1.4	Supervisor	2		2		
		1.5	Web services				600	
Engineering	Sewer Backup	2.1	Engineer	4		3		
		2.2	Web services				800	
	Electrical Breaks	2.3	Electrician	13	6			
		2.4	Supervisor	2	1			
		2.5	Web services				600	
	Hwy. repair	2.6	Field engineer	10		4		
		2.7	Contracts	4		4		
City hall totals				90	15	31	4,600	
IT Department	Public		Web Services					**30,000**
Site 2 - Operations (WAN)								
Operations	Clean-up prog.	3.1	Ops. Staff	4		2	1,000	
Operations totals				4		2	1,000	
Remote field offices (WAN)								
Site 3 - Freeberg	Inspection	4.1	Field engineer	40		30	*700*	
Site 4 - Willsberg	Inspection	4.1	Field engineer	30		20	*500*	
Field Offices	Inspection	4.2	Web Services				1,200	
Remote field office totals				70		50	1,200	
City Totals				164	15	83	6,800	30,000

Figure 9.14 **Year 1 user requirements.**

requirements for the second year of GIS in Rome. But first, you need to know where the additional users will be coming from and how they will connect to the GIS.

It has been decided that a central computing environment will be maintained in the city hall IT department through the second year of GIS in Rome, and GIS operations will be extended to five additional remote sites and mobile operations with access through the city's Internet connections. The system will implement a necessary firewall for the police department data (see figure 9.15). The city database is replicated in the police department so the department can use it in combination with their confidential data to serve their office staff and, through wireless connections, to serve their vehicles independently.

In year 2 the Operations building will be served from city hall over the available WAN connection, and the 911 emergency services will be located there in year 2. The city hall GIS data server hosts the emergency services data, to which emergency services desktop clients have terminal access only indirectly, through the city hall Windows Terminal Server (WTS) farm that runs their GIS viewer desktop applications. Emergency web services are supported out of city hall on the enterprise web server farm over the city Internet connection. Emergency vehicles will access the 911 web services over wireless Internet connections to the city hall web services. Remote sites 3 and 4 will be connected over the WAN, while sites 5, 6, 7, 8, and 9 will be connected to city hall through their Internet connection.

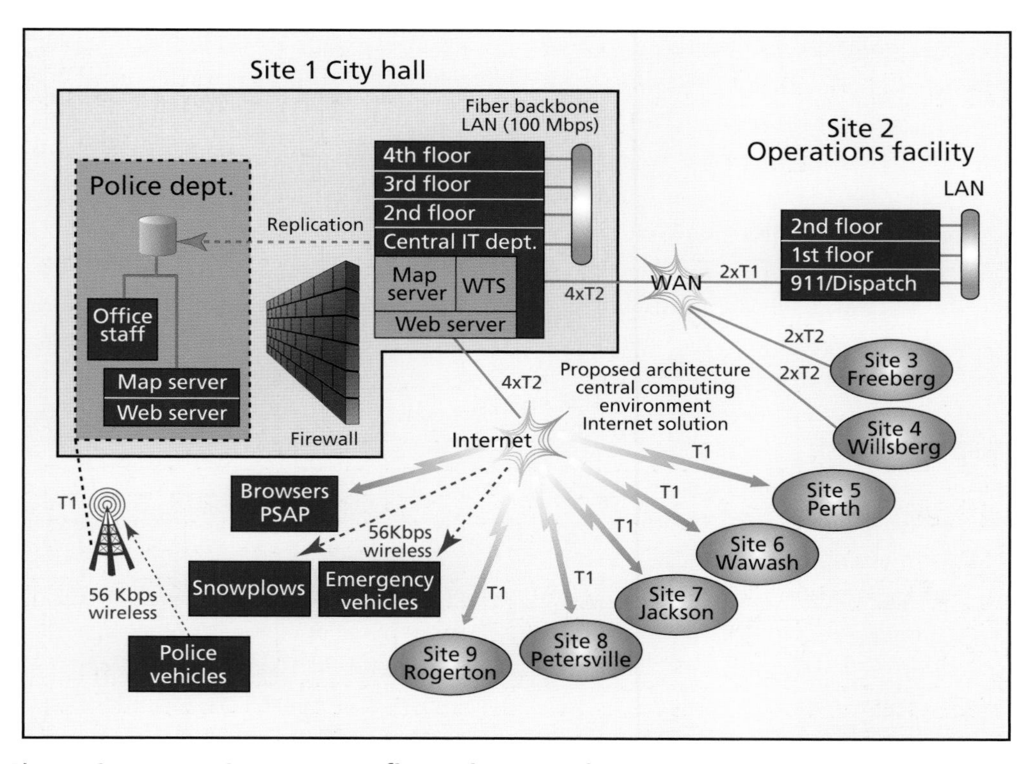

Figure 9.15 **Year 2 system configuration: user locations and network connectivity.**

The second year's requirements may include any known plans for expanding GIS throughout the organization, although the actual implementation may extend to future years. It is important to identify platform processing requirements over the hardware life cycle, so planning for future needs will avoid platform configuration shortfalls. Planning should be reevaluated on an annual basis to adjust for business and technology change.

Year 2 user requirements

In the same kind of table that you created for year 1, make your estimates for the second year. Assume new departments will need access to the GIS; remember that special requirements could emerge, such as the need for a firewall between the central computer and the police computer. Plan for operations to be extended to cover 911 emergency services and connections with vehicles; also, additional remote sites will need to be serviced.

With all this, you'll need to consider an extended architecture for the second year. What are your options for increasing system capabilities to meet Rome's increasing needs? All this is likely to confirm your opinion that, in planning the first year, you are wise to establish a system that is scalable, one that can grow as the organization grows. Just be sure that you are reflecting the objectives of the needs of the departments concerned.

The Operations building is currently served from city hall, and now the mayor has proposed that the 911 emergency services be located there in year 2. The city hall GIS data server hosts the emergency services data, to which emergency services desktop clients have terminal access only indirectly, through the city hall WTS that runs their low-complexity desktop applications. Emergency web services are supported out of city hall on the enterprise web server farm over the city Internet connection. Emergency vehicles will access the 911 web services over wireless Internet connections to the city hall web services. New remote sites 5, 6, 7, 8, and 9 will be connected to city hall over the data center Internet connections.

Several additional parts of the organization want to take fuller advantage of GIS capabilities, some for the first time, some in addition to current usage, and some to take advantage of software advances that did not exist previously.

The police department wants to take advantage of capabilities for replicating only selected parts of the city hall database (changes only) rather than incurring complete replication every night. It also wants to use GIS for crime analysis and introduce two new information products for police dispatching and optimal routing of patrol officers.

Several major improvements are required in the operations facility. The fire and ambulance dispatch needs a 911 in-vehicle routing display. Additionally, the snow clearing operations will be aided by GIS in their scheduling and vehicle tracking of one hundred snowplows using a Tracking Analyst product.

The engineering department intends to use GIS for work order optimizing in both the engineering and field operations departments. Five additional field offices (Perth, Wawash, Jackson, Petersville, and Rogerton), some with large staffs, have asked to be added to the GIS. In city hall, the business development department wants to use Esri Business Analyst to calculate suitability and emergency response times for a variety of proposed business sites. Both the business development department in city hall and the remote field offices expect an increase in the number of requests per hour on their web services.

The year 2 user requirements table (figure 9.16) summarizes the increase in the number of users in both the DeskEdit and DeskView workflows. Compare the increases in the peak number of requests per hour over the web from year 1 to year 2. Total local desktop users have more than doubled, the total remote desktop users increased by five times, and the number of web server requests per hour has tripled.

User requirements summary

Comparing the workloads of year 2 and year 1, the summary table (figure 9.17) makes the increases clear. In city hall, the engineering department has added five peak users. The business development department accounts for an additional eighteen peak users: those to be involved in site selection (eight), analysis of flood endangered land (eight), and the calculation of emergency response capability for potential business development (two).

The police department has assigned fifteen detectives as peak users in crime analysis; managing the police dispatch service will require an additional three peak users. In the operations facility, the fire and ambulance dispatch will need thirty more peak users, and scheduling for snow clearing adds another four.

The five new remote field offices (Perth, Wawash, Jackson, Petersville, and Rogerton) represent 182 peak users combined. When added to the peak users from the other remote sites (66 + 70 + 182 = 318), they make a

City of Rome - Year 2				Total Users	Peak User Workflow					
					Desktop (Users)			Server (req/hr)		
Department	Workflow	IPD	User type		DeskEdit	DeskView	Bus Anal	LocalMap	PublicMap	Mobile
Site 1 - City Hall (LAN)										
Planning	Zoning	1.0	Planner	25		15				
		1.1	Web services					2,600		
	Permits	1.2	Inspector	25		15				
		1.3	Appraiser	20	10					
		1.4	Supervisor	5	2					
		1.5	Web services					900		
Engineering	Sewer Backup	2.1	Engineer	5		3				
		2.2	Web services					1,000		
	Electrical Breaks	2.3	Electrician	13	6					
		2.4	Supervisor	2	1					
		2.5	Web services					1,900		
	Hwy. repair	2.6	Field engineer	11		7				
		2.7	Contracts	4		4				
	Work Orders	2.8	Managers	4		3				
	Work Reports	2.9	Field Units	10		2				
Business	Site Sel. and Natureserve	6.1	Planners	10			8			
Development	FMA Flood Zone, Serviced Land	6.2	Planners	10			8			
	Emergency Response, Time, etc.	6.3	Planners	2			2			
	Web Services	6.4						2,000		
City Hall LAN Totals				146	19	49	18	8,400		
IT Department	Public		Web Services					17,100	100,000	
Police (Firewall)	Patrol sched.	5.1	Admin.	10		3				
		5.5	Web services							3,600
	Crime analysis	5.2	Detectives	20	15					
	Spec. events	5.3	Traffic	10		3				
Remote Patrols	Police Dispatch	5.4	Traffic	10		3				
	Patrols/Routing	5.6	Patrol officers	20						*3,600*
Police Network Totals				70	15	9				3,600
Site 2 - Operations (WAN)										
Operations	Clean-up prog.	3.1	Ops. staff	4		2				
911	Response	3.2	Call takers	50		30				
		3.3	Web services					4,000		
Remote vehicles	Fire and Ambalance Dispatch	3.4	Schedulers	30		30				
	Routing	3.5	Drivers	30						
Snow Clearing	Scheduling	3.6	Engineers	4		4				
	Snow Plows	3.7	Drivers	100						
Operations totals				218	0	66		4,000		
Remote field offices (WAN)										
Site 3 - Freeberg	Inspection	4.1	Field engineer	45		30		*700*		
Site 4 - Willsberg	Inspection	4.1	Field engineer	60		40		*1,000*		
WAN Field Offices	Inspection	4.2	Web Services					*1,700*		
Remote field office (WAN) totals				105	0	70		1,700		
Remote field offices (Internet)										
Site 5 - Perth	Inspection	4.3	Field engineer	10		2		*100*		
Site 6 - Wawash	Inspection	4.3	Field engineer	50		40		*500*		
Site 7 - Jackson	Inspection	4.3	Field engineer	60		20		*400*		
Site 8 - Petersville	Inspection	4.3	Field engineer	80		60		*1,000*		
Site 9 - Rogerton	Inspection	4.3	Field engineer	80		60		*1,000*		
Internet Field Offices	Inspection	4.2	Web Services					*3,000*		
Remote field office (Internet) totals				280	0	182		3,000		
City totals (excluding Police private network)				749	19	367	18	17,100	100,000	

Figure 9.16 **Year 2 user requirements.**

total of 318 remote desktop users accessing the Windows Terminal Server, along with 24 users accessing the police network production server, and an overall total of 404 desktop users (19 DeskEdit + 367 DeskView + 18 Business Analyst) on the city enterprise data server. The anticipated increase in web requests per hour (including 2,000 from the business development department and 4,000 from the remote field offices) result in a new total of 17,100 requests per hour for local web mapping services. Public mapping services are expected to peak at up to 100,000 requests per hour. The police network will provide mobile synchronization services to the police patrols, with peak levels estimated at 3,600 web requests per hour (20 police cars each synchronized 3 times each minute).

Figure 9.17 identifies peak usage for each user department, by location, for each of the two years. It also represents peak user loads for year 1 and 2 based on requirements identified by each department manager. These are the same user loads identified in the earlier year 1 and year 2 user requirements tables, but providing them in a compressed format here makes sense for two reasons: (1) it's simply easier for managers to sign off on the requirements if they can compare estimates for the two planned years side by side, and (2) you will be using this summary to complete the network bandwidth suitability and platform loads analyses.

City of Rome workflow requirements analysis

The IPDs and MIDL discussed in chapter 6 provide a foundation for establishing workflow use cases required to complete the system architecture design. Based on results from the user workflow needs assessment, City of Rome will require the following nine separate user workflows:

- GIS desktop editor (GISDeskEditor)
- GIS desktop viewer (GISDeskView)
- GIS desktop Business Analyst (GISDeskBA)
- Remote GIS desktop viewer (RemoteGISView)

City of Rome	Year 1					Year 2						
		Peak Usage					Peak Usage					
	Total	GIS Desktop		Server		Total	GIS Desktop (users)			Server (Req/hr)		
Department	Users	DeskEdit	DeskView	LocalMap	PublicMap	Users	DeskEdit	DeskView	Bus Anal	LocalMap	PublicMap	Mobile
Site 1 - City Hall												
Planning	57	8	20	3,200		75	12	30		3,500		
Engineering	33	7	11	1,400		49	7	19		2,900		
Business Development						22			18	2,000		
City Hall LAN Totals	90	15	31	4,600		146	19	49	18	8,400		
IT Department	Web Services			6,800	30,000	Web Services				17,100	100,000	
Firewall												
Police						40	15	6				3,600
Dispatch						10		3				
Remote Patrols						20						3,600
Police Network Totals						70	15	9				3,600
Operations and Remote Field Offices (WAN)												
Site 2 - Operations												
Operations	4		2	1,000		54		32		4,000		
Remote Vehicles						60		30				
Snow Clearing						104		4				
Operations totals	4		2	1,000		218		66		4,000		
Site 3 - Freeberg	40		30	700		45		30		700		
Site 4 - Willsberg	30		20	500		60		40		1,000		
Remote field office (WAN) totals	70		50	1,200		105		70		1,700		
Remote field offices (Internet)												
Site 5 - Perth						10		2		100		
Site 6 - Wawash						50		40		500		
Site 7 - Jackson						60		20		400		
Site 8 - Petersville						80		60		1,000		
Site 9 - Rogerton						80		60		1,000		
Remote field office (Internet) totals								182		3,000		
City totals (excluding Police private network)	164	15	83	6,800	30,000	469	19	367	18	17,100	100,000	

Figure 9.17 **User requirements summary.**

- GIS server internal web mapping services (WebInternal)
- GIS server public web mapping services (WebPublic)
- Batch process for background system administration (BatchAdmin)
- Mobile client (MobileClient)
- Mobile synchronization service (MobileService)

Custom City of Rome workflow performance targets (maximum acceptable processing time) will be generated following the methodology identified in appendix D. City of Rome chose to use online community basemaps for their workflows. Figure 9.18 identifies the workflow recipe used to generate the custom City of Rome workflows.

Figure 9.18 lists workflow abbreviations and their descriptions, also known as workflow recipes. The workflow recipe is used to generate custom performance targets for client traffic and software services times for each workflow. The custom targets provided in figure 9.19 will be used to complete the system architecture design.

Workflow	Workflow Description
=== City of Rome workflows ===	=== City of Rome workflows ===
GISDeskEditor	AGD101 wkstn MXD 50%Dyn Med 10x7 Feature +$S
GISDeskView	AGD101 wkstn MXD 50%Dyn Lite 10x7 Feature +$S
GISDeskBA	AGD101 wkstn MXD 50%Dyn Med 10x7 Feature +$S
RemoteGISView	AGD101 Citrix MXD V 50%Dyn Med 10x7 ICA +$S
WebInternal	AGS101 REST MSD V 50%Dyn Med 10x7 PNG24 +$S
WebPublic	AGS101 REST MSD V 50%Dyn Lite 10x7 PNG24 +$S
BatchAdmin	AGS101 REST MSD R 100%Dyn Med 10x7 JPEG (no client)
MobileClient	Client device load only: AGD101 wkstn MXD 100%Dyn Lite 4x3 Feature
MobileService	AGS101 SOAP MXD V 10%Dyn Lite 4x3 Feature

Figure 9.18 **City of Rome workflow performance recipes.**

Workflow	Software Component Service Times							
		Arc12 baseline			SRint06/core = 45.0			Software Service Time
	Client Traffic	Design Model Metrics			Database			
		Client		Web	SOC	Data	Data	
=== City of Rome workflows ===	Mbpd	Client	Citrix	Web	SOC	SDE	DBMS	Total
GISDeskEditor	5.500	0.169				0.021	0.021	0.211
GISDeskView	3.000	0.085				0.011	0.011	0.107
GISDeskBA	5.500	0.169				0.021	0.021	0.211
RemoteGISView	0.710	0.036	0.169			0.021	0.021	0.247
WebInternal	1.700	0.036		0.015	0.075	0.014	0.014	0.154
WebPublic	1.700	0.036		0.015	0.038	0.007	0.007	0.103
BatchAdmin				0.012	0.116	0.029	0.029	0.186
MobileClient		0.166						0.166
MobileService	0.076	0.036		0.023	0.009	0.001	0.001	0.070

Figure 9.19 **City of Rome custom workflow performance targets.**

Year 1 bandwidth suitability analysis

To assess whether the available bandwidth is adequate for the use proposed, we'll use the traffic per display identified in the custom workflow performance targets along with the peak user loads to estimate peak network traffic for each workflow. The GIS peak traffic projections along with our bandwidth recommendations will be provided to the network administrator to support appropriate planning of enterprise network bandwidth capacity needs (GIS traffic is only part of the overall enterprise traffic over the WAN connections).

Figure 9.20 shows the logic for completing the network bandwidth suitability analysis. The spreadsheet is a representation of the Rome year 1 user requirements already identified (see figure 9.17). Data center network connections are represented by the gray rows (LAN backbone, WAN connection, Internet connection). Remote sites are represented by the green rows (Operations, Freeberg, Willsberg). Public web services connect over the data center Internet connection. Connection bandwidth for each network is identified in the Bandwidth column.

In figure 9.20, you can see that city hall has two desktop workflows (15 desktop editors and 31 desktop viewers) plus one batch process for use by system administrators. One remote desktop viewer workflow is located at each of the remote WAN sites (2 peak users at Operations, 30 at Freeberg, and 20 at Willsberg for a total of 52 concurrent remote desktop viewers). Internal web services are used by local and WAN clients with peak throughput load expressed as peak requests per hour (4,600 on the local network, 1,000 at Operations, 500 at Freeberg, 700 at Willsberg). Public web services are published over the data center Internet connection (30,000 map requests per hour).

City of Rome Year 1	User Environment					Bandwidth Mbps	NW %Cap
Types of Workflows	Peak Concurrent		DPM/TPM		Network		Traffic
Standard	Users	TPH	per Client	Total	Mbps		Mbpd
LAN_Local Clients		Services	47 Clients	LAN = 31.4 Mbps		100	31%
GISDeskEditor	15		10.00	150	13.750		5.500
GISDeskView	31		10.00	310	15.500		3.000
WebInternal		4,600	6.00	77	2.172		1.700
BatchAdmin	1		6.00	6			
WAN_Clients			52 Clients	WAN = 7.2 Mbps		1.5	479%
Site 2_Operations			2 Clients	Trafic = 0.7 Mbps		1.5	47%
RemoteGISView	2		10.00	20	0.237		0.710
WebInternal		1,000	6.00	17	0.472		1.700
Site 3_Freeberg			30 Clients	Traffic = 3.8 Mbps		1.5	252%
RemoteGISView	30		10.00	300	3.550		0.710
WebInternal		500	6.00	8	0.236		1.700
Site 4_Willsberg			20 Clients	Traffic = 2.7 Mbps		1.5	180%
RemoteGISView	20		10.00	200	2.367		0.710
WebInternal		700	6.00	12	0.331		1.700
Internet_Clients				Internet = 14.2 Mbps		1.5	944%
Public_Web Services				Traffic = 14.2 Mbps		1.5	944%
WebPublic		30,000	6.00	500	14.167		1.700

Figure 9.20 **Rome year 1: Network suitability analysis.**

The network traffic for each workflow can be calculated from the peak display throughput (displays per minute, or DPM) and the traffic per display (Mbpd): multiply displays per minute by megabits per display and divide by 60 seconds—the result is megabits per second traffic for the given workflow. You can then sum the workflow traffic results to estimate traffic passing through the remote site and data center WAN and Internet connections. The spreadsheet identifies network communication segments that have more traffic than bandwidth in red—these workflows will experience severe performance problems if the network bandwidth is not upgraded to handle peak workflow traffic loads. The remote site and data center network connection bandwidth should be roughly twice the projected traffic to avoid network contention delays.

Peak workflow display throughput (the "Total" column) is calculated by multiplying the peak concurrent users by the user productivity (the "per client" column). User productivity can vary for different workflow use cases. A single custom workflow can be used to support cases with different productivity by listing them as separate rows in the user requirements analysis. For planning purposes, a GIS desktop user has a maximum productivity of 10 displays per minute. Web service transaction loads can be identified by peak hour throughput rates (transactions per hour) or as peak concurrent users times user productivity. Web users are less productive with a maximum productivity of 6 displays per minute. When identifying peak web throughput requirements, the network traffic, productivity, and throughput cells can be shown in red if calculated throughput is not supported with available bandwidth.

The network suitability analysis identifies peak network connection traffic between sites, which can then be used to identify network connection bandwidth requirements. Network bandwidth should be roughly twice the projected peak traffic loads for each connection; user display performance can deteriorate rapidly as traffic exceeds 50 percent utilization rates due to traffic contention. (Similar to rush hour traffic on a freeway, network traffic can come to a standstill if there is not enough bandwidth.) Network upgrade recommendations for Rome year 1 are provided in figure 9.21.

City hall will need to update the Data Center WAN and Internet connections.

- WAN: Network traffic of 7.2 Mbps. Upgrade bandwidth from T1 (1.5 Mbps) to 4xT2 (24 Mbps).
- Internet: Network traffic of 14.2 Mbps. Upgrade bandwidth from TI (1.5 Mbps) to 4xT2 (24 Mbps). Note: Public Internet is not a single site and can be represented by the Data Center bandwidth.

Two WAN remote sites will require network upgrades.

- Site 2_Operations: Site traffic of 0.7 Mbps. Upgrade bandwidth from T1 (1.5 Mbps) to 2xT1 (3 Mbps).
- Site 3_Freeberg: Site traffic of 3.8 Mbps. Upgrade bandwidth from T1 (1.5 Mbps) to 2xT2 (12 Mbps).
- Site 4_Willsberg: Site traffic of 2.7 Mbps. Upgrade bandwidth from T1 (1.5 Mbps) to 2xT2 (12 Mbps).

Figure 9.21 **Rome year 1: Network upgrade recommendations.**

Year 1 workflow platform loads analysis

The first step in the platform loads analysis is to identify the workflow software configuration. Figure 9.19 identified the custom baseline software component service times for the City of Rome workflows. Each of the workflow software components, other than the client software, are installed on server platforms in the data center.

Figure 9.22 shows the City of Rome year 1 workflows in the first column. Software components (client, WTS Citrix, web, SOC, DBMS) for each of these workflows are identified in the software configuration section of the table. The server the software is installed on is identified below each software component (default configuration is identified on the LAN row—each workflow is assigned to the default platform for the year 1 install except for the WebPublic workflow). WebPublic workflow web and SOC software are installed on the WebPub server. Installing the public web services on a separate application server platform protects inbound services from being compromised from external web service access.

The combined server platform processing loads are identified in the platform service time columns. For example, consider the WebInternal workflow. Figure 9.19 provides the City of Rome custom service times (processing time per display), which includes web (0.015 sec), SOC (0.075 sec), SDE (0.014 sec), and DBMS (0.014 sec) software components that run in the computer room. The web and SOC components are installed on the WebIn server (SDE default is a direct connect configuration, which is executed by the SOC). Back on figure 9.22, you see the total WebIn server platform service time is rounded to 0.104 sec (0.015 + 0.075 + 0.014). The DBMS software (0.014 sec) is installed on the DBMS server. The client browser service time is 0.036 sec. The platform service times shown in the table will be used to complete the integrated workflow platform loads analysis.

The next step is to identify the total baseline software processing time for each of the installed data center servers. Figure 9.23 shows how to complete this analysis. The user workflows are configured in their user site locations as described in the network suitability analysis (figure 9.20). Peak processing load for each workflow is calculated by multiplying the peak throughput (displays per minute) by the appropriate platform service time, and then the total workflow processing times are collected on the installed platforms.

For example, in figure 9.23, find the RemoteGISView workflow for the Freeberg site. It has a peak "total" throughput of 300 displays per minute with installed baseline service times for the WTS (0.190 sec) and DBMS (0.021 sec) platforms (client time is supported by the user desktop). Total baseline processing time for the WTS server is 57.0 seconds (300 × 0.190), and 6.3 seconds (300 × 0.021) is the rounded number for the DBMS server.

City of Rome Year 1 Types of Workflows	Software Configuration					Platform Service Times Arc12 4 core (1 chip) Baseline				
	Desktop		Web	Server	Data Source					
Standard	Client	Citrix	Web	SOC	DBMS	Client	WTS	WebIn	WebPub	DBMS
LAN_Local Clients	Client	WTS	WebIn	WebIn	DBMS	Platform Service Times (seconds)				
GISDeskEditor	Default				Default	0.190				0.021
GISDeskView	Default				Default	0.096				0.011
WebInternal	Default		Default	Default	Default	0.036		0.104		0.014
BatchAdmin			Default	Default	Default			0.157		0.029
RemoteGISView	Default	Default			Default	0.036	0.190			0.021
WebPublic	Default		WebPub	WebPub	Default	0.036			0.060	0.007

Figure 9.22 **Rome year 1 platform configuration.**

Once all of the workflow processing times are collected for each platform, the processing times can be added together to get the total platform baseline processing load. For example, the total WebPub map server processing load is 30.0 seconds.

This completes the baseline platform loads analysis. We now have the total baseline processing time for each installed data center server platform. Figure 9.23 shows how these times are passed on to complete our final platform configuration analysis.

The computer center configuration includes four server tiers (WTS platform tier, internal web mapping tier, public web mapping tier, and database tier). The total baseline processing time for each server tier has been identified and carried over to the platform selection table. You can now select the platform configuration you wish to use in your data center. The highest performance server platforms will provide the best user experience. The Xeon E5-2637 platform was selected for the City of Rome minimum physical server platform solution shown in figure 9.23.

Three different platform solutions will be considered for City of Rome year 1 deployment:

- Minimum physical server configuration
- High-availability physical server configuration
- High-availability virtual server configuration

Hardware pricing will be calculated for each configuration candidate.

City of Rome Year 1	User Environment					Bandwidth Mbps	NW %Cap	Platform Service Times Arc12 4 core (1 chip) Baseline					Total baseline platform loads								
Types of Workflows	Peak Concurrent		DPM/TPM		Network		Traffic						Client	WTS		WebIn		WebPub		DBMS	
Standard	Users	TPH	per Client	Total	Mbps		Mbpd	Client	WTS	WebIn	WebPub	DBMS	Users	Users	Sec	Users	Sec	Users	Sec	Users	Sec
LAN_Local Clients		Services	47 Clients	LAN = 31.4 Mbps		100	31%	Platform Service Times (seconds)													
GISDeskEditor	15		10.00	150	13.750		5.500	0.190				0.021	15.0							15.0	3.2
GISDeskView	31		10.00	310	15.500		3.000	0.096				0.011	31.0							31.0	3.4
WebInternal		4,600	6.00	77	2.172		1.700	0.036		0.104		0.014	12.8			12.8	8.0			12.8	1.1
BatchAdmin	1		331.48	331						0.157		0.029				1.0	52.0			1.0	9.6
WAN_Clients			52 Clients	WAN = 7.2 Mbps		24	30%														
Site 2_Operations			2 Clients	Trafic = 0.7 Mbps		3	24%														
RemoteGISView	2		10.00	20	0.237		0.710	0.036	0.190			0.021	2.0	2.0	3.8					2.0	0.4
WebInternal		1,000	6.00	17	0.472		1.700	0.036		0.104		0.014	2.8			2.8	1.7			2.8	0.2
Site 3_Freeberg			30 Clients	Trafic = 3.8 Mbps		12	32%														
RemoteGISView	30		10.00	300	3.550		0.710	0.036	0.190			0.021	30.0	30.0	57.0					30.0	6.3
WebInternal		500	6.00	8	0.236		1.700	0.036		0.104		0.014	1.4			1.4	0.9			1.4	0.1
Site 4_Willsberg			20 Clients	Trafic = 2.7 Mbps		12	22%														
RemoteGISView	20		10.00	200	2.367		0.710	0.036	0.190			0.021	20.0	20.0	38.0					20.0	4.2
WebInternal		700	6.00	12	0.331		1.700	0.036		0.104		0.014	1.9			1.9	1.2			1.9	0.2
Internet_Clients				Internet = 14.2 Mbps		24	59%														
Public_Web Services				Trafic = 14.2 Mbps		24	59%														
WebPublic		30,000	6.00	500	14.167		1.700	0.036			0.060	0.007	83.3					83.3	30.0	83.3	3.5
						1.5															
Total Throughput	99	36,800											200.2	52.0	98.8	19.9	63.8	83.3	30.0	201.2	32.2

Server Platform tier	Baseline sec	Platform Selection Platform Configuration	SRint2006 per Core
WTS: WTS Platform Tier	98.8	Xeon E5-2637 4 core (2 chip) 3000 MHz	46.3
WebIn: Internal Mapping Tier	63.8	Xeon E5-2637 4 core (2 chip) 3000 MHz	46.3
WebPub: Public Mapping Tier	30.0	Xeon E5-2637 4 core (2 chip) 3000 MHz	46.3
DBMS: Database Tier	32.2	Xeon E5-2637 4 core (2 chip) 3000 MHz	46.3

Figure 9.23 **Rome year 1 physical server workflow platform loads analysis.**

Hardware pricing model

Figure 9.24 shows a hardware pricing model identifying the available platform selections and the vendor pricing summary from the City of Rome procurement office, along with the vendor published benchmark performance (SPECrate_int2006 baseline).

The SPECrate_int2006 performance benchmark published by the vendor provides relative platform performance information we will need to complete our analysis.

Memory requirements must be identified to complete the platform selection in figure 9.25. Memory requirement guidelines for Windows Terminal Server and database platforms are based on peak concurrent number of user sessions and database connections. Memory requirements can be estimated by multiplying 4 GB by the platform SPECrate_int2006 baseline throughput divided by 14. Web and map server memory requirement guidelines are based on maximum number of deployed service instances. Memory can be estimated by multiplying 3 GB by the number of platform core.

FOCUS

Alternative system design strategies

There are several alternative system design strategies you need to consider before you can adequately choose the right platform architecture:

- **High availability requirements** Most enterprise environments require the system to continue to operate with loss of a server platform. For the Windows Terminal Server platform tier, an N+1 configuration (one extra server over your required platforms) is recommended for high availability. For web and mapping servers, a minimum of two servers should be included in the web mapping tier. The database tier should include one failover server to complete the production environment.
- **Development, test, and staging servers** Most enterprise environments maintain additional platforms configured with the same operating system as the production environment for development, testing, and staging. These servers do not have performance and capacity requirements, so often you can use older server platforms for these environments.
- **Virtual server environments** Many enterprise environments today are using virtual server technology to consolidate their data center environment. Virtual servers offer many administrative advantages and can contribute to a more stable and adaptive server environment. Multiple virtual servers can be hosted on a single physical server, enabling better utilization of the current high-capacity Intel Xeon X5677 8-core server environments. Performance of the virtual servers will depend on their host physical server environments, and you can expect some processing overhead (you should include 50 percent extra processing overhead for 4-core virtual server environments and 80 percent extra processing overhead for 8-core servers).
- **Cloud hosting services** Software as a service, platform as a service, and infrastructure as a service are alternative deployment options you might consider. Amazon infrastructure service offerings can be used to reduce City of Rome data center platform processing and traffic loads, and enable rapid platform deployment and elastic on-demand service capacity for future deployments.

	Hardware Platforms	Memory	SPECrate_int2006		Price
			Throughput	per core	
4 core	Xeon E5-2637 4 core (2 chip) 3000 MHz	12 GB RAM	185	46.3	$13,485
	Xeon E5-2637 4 core (2 chip) 3000 MHz	48 GB RAM	185	46.3	$14,385
8 core	Xeon E5-2643 8 core (2 chip) 3300 MHz	24 GB RAM	358	44.8	$13,887
	Xeon E5-2643 8 core (2 chip) 3300 MHz	96 GB RAM	358	44.8	$15,687
12 core	Xeon E5-2667 12 core (2 chip) 2900 MHz	36 GB RAM	501	41.8	$14,885
	Xeon E5-2667 12 core (2 chip) 2900 MHz	96 GB RAM	501	41.8	$16,385
	Xeon E5-2667 12 core (2 chip) 2900 MHz	128 GB RAM	501	41.8	$17,185
16 core	Xeon E5-2690 16 core (2 chip) 2900 MHz	48 GB RAM	662	41.4	$16,025
	Xeon E5-2690 16 core (2 chip) 2900 MHz	96 GB RAM	501	41.8	$17,225
	Xeon E5-2690 16 core (2 chip) 2900 MHz	128 GB RAM	501	41.8	$18,025
	Xeon E5-2690 16 core (2 chip) 2900 MHz	192 GB RAM	662	41.4	$19,625

Figure 9.24 **Hardware pricing model (US dollars, 2012).**

Year 1 minimum physical server configuration

Figure 9.25 identifies steps required to complete the platform sizing and pricing analysis. The platform baseline processing time is converted to equivalent processing time on the selected platform by multiplying the processing load (baseline sec) by the baseline platform SPEC benchmark (platform baseline SRint2006 = 45), then dividing the result by the SRint2006 per-core benchmark results of the selected platform. This formula provides the adjusted platform processing loads.

User display response times will increase as platform CPU utilization rates approach 80 to 90 percent. To achieve good performance, you should add additional platforms (rollover) whenever processing loads exceed 80 percent platform utilization. The number of platform core needed to stay below 80 percent utilization is provided in the "Required Core" column of figure 9.25 (adjusted processing time divided by percent rollover multiplied by 60 seconds). The total number of required platforms (nodes) for each platform tier is calculated by taking the required core and dividing by the selected platform core, rounded up to the nearest whole number. RAM per node is calculated based on memory requirement guidelines for each platform tier, then the next higher available memory is selected from our hardware procurement list (figure 9.24), Platform utilization is calculated by taking the required core and dividing by the total number of platform core.

For example, the WTS Platform tier has a total baseline processing load of 98.8 seconds. The Xeon E5-2637 4 core server was the selected platform configuration. The adjusted processing time was 96.1 seconds (98.8 × 45/46.3). Divide 96.1 by 60 to identify the required processor core of 1.6. This server has 4 processor cores, so 1 server node will satisfy the processing requirements. Server utilization is rounded off to 40 percent (1.6 core / 4 core); additional server nodes will be included if utilization is above the 80 percent rollover value. Price per node is $14,385. Total cost for purchasing the required Windows Terminal Servers is $14,385 (1 node × $14,385).

The City of Rome Web Mapping and DBMS platform loads are relatively light, so it makes sense to purchase Xeon E5-2637 4-core servers for those tiers. The platform node cost is provided on your platform pricing list (figure 9.24) based on the selected platform. Multiply the platform price by the required number of platform nodes to identify total cost. It is important to review your future system load requirements (year 2) to make sure your initial purchase will satisfy needs over the hardware life cycle (normally 3 years). The total hardware cost for the minimum physical server configuration in figure 9.25 is $55,740.

Batch process workflow platform loads analysis

One final note about the workflow platform loads analysis shown in figure 9.23. Batch process workflows are quite common in GIS (often referred to as batch geoprocessing services). Any procedure that takes more than 1 minute to execute is best configured as a background batch process (not part of the user workflow). Batch processes are normally sequential operations, which means that the program instructions are executed one after the other throughout the process run. Each batch process will take advantage of only one platform processor core at a time, which means that a single batch process will leverage only one processor core on each installed platform tier.

Batch processes must be handled differently than typical user workflows. The productivity (displays per minute) of each batch process will depend on the processor core performance of the selected platform. Productivity for the batch process is determined by dividing the productivity interval (60 seconds) by the display response time. Display is a work unit that represents one iteration of a batch process; a work unit can be an arbitrary unit of processing time (the total batch process can be divided into many work units).

To compute the batch display response time, the baseline service times must first be adjusted to represent the selected platform service times (Xeon E5-2637 platforms were selected for the City of Rome physical server configurations). The baseline service times are identified in the BatchAdmin workflow row in figure 9.22 (WebIn = 0.157, DBMS = 0.029). The baseline service times are converted to selected platform service times (WebIn

Server Platform tier	Baseline sec	Platform Selection Platform Configuration	SRint2006 per Core	Adjusted sec
WTS: WTS Platform Tier	98.8	Xeon E5-2637 4 core (2 chip) 3000 MHz	46.3	96.1
WebIn: Internal Mapping Tier	63.8	Xeon E5-2637 4 core (2 chip) 3000 MHz	46.3	62.1
WebPub: Public Mapping Tier	30.0	Xeon E5-2637 4 core (2 chip) 3000 MHz	46.3	29.2
DBMS: Database Tier	32.2	Xeon E5-2637 4 core (2 chip) 3000 MHz	46.3	31.3

Server Platform tier	Adjusted sec	Percent Rollover	Required Core	Platform Core	Platform Nodes	RAM per Node	Node Cost	Total Cost	Platform Utilization
WTS: WTS Platform Tier	96.1	80%	1.6	4	1	48 GB	$14,385	$14,385	40%
WebIn: Internal Mapping Tier	62.1	80%	1.0	4	1	12 GB	$13,485	$13,485	26%
WebPub: Public Mapping Tie	29.2	80%	0.5	4	1	12 GB	$13,485	$13,485	12%
DBMS: Database Tier	31.3	80%	0.5	4	1	48 GB	$14,385	$14,385	13%
Storage pricing estimates: Volume of data gigabtes + 50% (data indexing) x $ per GB at required RAID level (see data storage estimates)									
							Total cost =	$55,740	

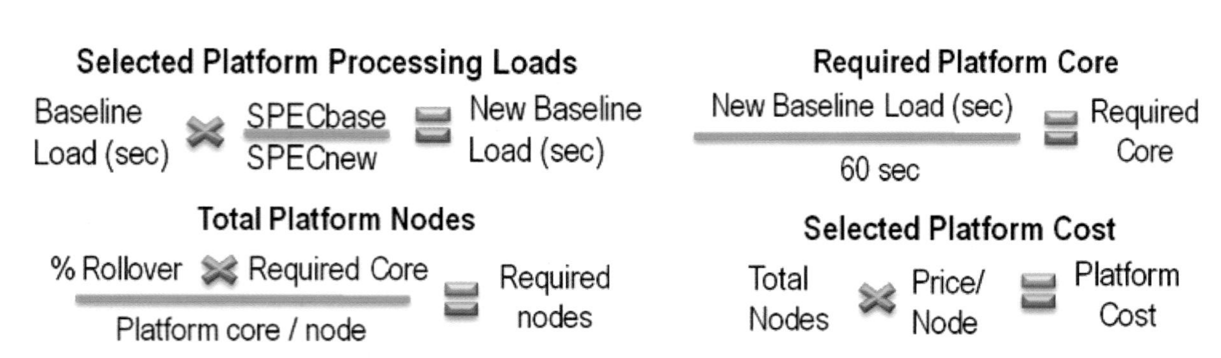

Figure 9.25 **Rome year 1 minimum physical server platform selection and pricing.**

= 0.153, DBMS = 0.028) by multiplying the baseline service time by the baseline platform SPEC benchmark (platform baseline SRint2006 = 45), then dividing the result by the SRint2006 per-core benchmark results of the selected platform (46.3). Batch process cycle time is 0.181 sec (0.153 + 0.028). Batch productivity is 331.5 displays per minute (60/0.181). For a single batch process, the baseline platform processing loads would be the baseline platform service times multiplied by the productivity (WebIn = 52 sec, DBMS = 9.6 sec).

For many batch processes you may not be sure what to use for display service times. Since the batch process will consume available processor core resources, the platform load will be the same as long as you provide the proper service time profile. For example, the batch process explained previously spends about 84.4 percent of the processing on the map server and only 15.6 percent on the DBMS. You could represent this same batch process with baseline service times that match this profile (WebIn = 0.844, DBMS = 0.156). Next, adjust these service times for the selected platform (WebIn = 0.821, DBMS = 0.152). Batch process cycle time is 0.973 sec. Productivity is 61.67 DPM. Platform processing loads are the same as the above example (Map = 52 sec, DBMS = 9.6 sec).

Year 1 high-availability physical server configuration

The year 1 high-availability physical server configuration will require a minimum of two servers for each platform tier. The Xeon E5-2637 4-core servers and the loads analysis in figure 9.23 can be used to complete the sizing analysis. The complete year 1 high-availability physical server platform selection and pricing analysis is described in appendix E.

Twice as many physical platforms will be required for the high-availability configuration.

Total hardware cost for the high-availability server configuration is $111,480.

Year 1 high-availability virtual server configuration

The high-availability virtual server configuration will deploy multiple virtual server machines on selected physical host machines. The City of Rome will use the high-capacity Xeon E5-2667 12-core (2-chip) 2900 MHz platforms for their virtual server host machines. The selected host platform will determine the processing performance for the virtual server machines. The complete year 1 high-availability virtual server workflow platform loads, platform selection and pricing analysis are described in appendix E.

Total hardware server cost includes two host server platforms, with each processor chip configured to host three virtual server machines. Total hardware cost (including the two host platforms and the additional virtual server machines) is $42,370.

Year 2 bandwidth suitability analysis

To determine proper bandwidth for year 2, we will follow the same methodology introduced with the year 1 bandwidth suitability analysis. Refer to the user requirements summary in figure 9.17 to identify year 2 user requirements updates. The year 2 custom workflow performance targets were already identified in figure 9.19.

Figure 9.26 shows the year 2 network suitability analysis. We added the new business development workflow to the city hall LAN (GISDeskBA). Additionally, we included five additional remote sites connecting to the data center over the Internet as shown in the year 2 system configuration drawing in figure 9.15 (each remote site includes remote GIS viewer terminal client and web internal workflows). Finally, we updated peak concurrent users based on information collected from the user requirements summary (figure 9.17).

Once all of the user workflows are configured on the proper site locations, you can follow the methodology discussed for year 1 to complete the network suitability analysis. The GIS has expanded significantly from year 1 to year 2, which has significantly increased demand on the City of Rome network infrastructure. Once you complete the year 2 workflow throughput and traffic calculations, you will identify upgrade requirements for the site network connections. Here, red cells indicate traffic that exceeds available bandwidth. Yellow network cells indicate traffic over 50 percent of the available bandwidth. You are now ready to visit with the network administrator to discuss required infrastructure upgrades. Red cells in

City of Rome Year 2						Bandwidth	NW
	User Environment					Mbps	%Cap
Types of Workflows	Peak Concurrent		DPM/TPM		Network		Traffic
Standard	Users	TPH	per Client	Total	Mbps		Mbpd
LAN_Local Clients		Services	87 Clients	LAN = 62.4 Mbps		100	62%
GISDeskEditor	19		10.00	190	17.417		5.500
GISDeskView	49		10.00	490	24.500		3.000
GISDeskBA	18		10.00	180	16.500		5.500
WebInternal		8,400	6.00	140	3.967		1.700
BatchAdmin	1		6.00	6			
WAN_Clients			136 Clients	WAN = 18.8 Mbps		24	78%
Site 2_Operations			66 Clients	Traffic = 9.7 Mbps		3	323%
RemoteGISView	66		10.00	660	7.810		0.710
WebInternal		4,000	6.00	67	1.889		1.700
Site 3_Freeberg			30 Clients	Traffic = 3.9 Mbps		12	32%
RemoteGISView	30		10.00	300	3.550		0.710
WebInternal		700	6.00	12	0.331		1.700
Site 4_Willsberg			40 Clients	Traffic = 5.2 Mbps		12	43%
RemoteGISView	40		10.00	400	4.733		0.710
WebInternal		1,000	6.00	17	0.472		1.700
Internet_Clients			182 Clients	Internet = 70.2 Mbps		24	292%
Public_Web Services				Traffic = 47.2 Mbps		24	197%
WebPublic		100,000	6.00	1,667	47.222		1.700
Site 5_Perth			2 Clients	Traffic = 0.3 Mbps		1.5	19%
RemoteGISView	2		10.00	20	0.237		0.710
WebInternal		100	6.00	2	0.047		1.700
Site 6_Wawash			40 Clients	Traffic = 5.0 Mbps		1.5	331%
RemoteGISView	40		10.00	400	4.733		0.710
WebInternal		500	6.00	8	0.236		1.700
Site 7_Jackson			20 Clients	Traffic = 2.6 Mbps		1.5	170%
RemoteGISView	20		10.00	200	2.367		0.710
WebInternal		400	6.00	7	0.189		1.700
Site 8_Petersville			60 Clients	Traffic = 7.6 Mbps		1.5	505%
RemoteGISView	60		10.00	600	7.100		0.710
WebInternal		1,000	6.00	17	0.472		1.700
Site 9_Rogerton			60 Clients	Traffic = 7.6 Mbps		1.5	505%
RemoteGISView	60		10.00	600	7.100		0.710
WebInternal		1,000	6.00	17	0.472		1.700

Figure 9.26 **Rome year 2: Network suitability analysis.**

the client displays per minute (DPM) and peak concurrent throughput columns represent the need to increase network bandwidth to avoid reduced user productivity and resolve network traffic constraints (the required use productivity cannot be maintained with the identified network performance bottlenecks).

Network upgrade recommendations are provided in figure 9.27.

Identifying and updating infrastructure budgets to handle the expected increase in GIS traffic is essential for successful year 2 deployment. If network upgrades are not in the budget, you risk facing performance problems that may not surface until after system deployment, when peak performance is critical to support business operations. The cost for not getting it right during the design can lead to painful results.

Year 2 workflow platform loads analysis

The next step is to update the workflow software installation. City of Rome has decided to keep the same four-platform tier centralized workflow software component configuration used for year 1. A desktop business analysis workflow is created for the business development department. This is a medium-complexity workflow with a baseline DBMS service time of 0.021 sec. The year 2 software configuration is shown in figure 9.28.

Figure 9.29 shows the city network high-availability virtual server platform loads analysis for year 2 deployments. WTS baseline platform server loads increased by over 611 percent (604.2/98.8), web mapping server loads increased by over 208 percent (168.8/80.9), and DBMS loads increased by over 344 percent (102.8/29.8). The BatchAdmin productivity is established following the

City hall will need to update the Data Center LAN backbone, WAN, and Internet connections.

- LAN: Network traffic of 62.4 Mbps. Upgrade backbone from 100 Mbps to 1 Gbps.
- WAN: Network traffic of 18.8 Mbps. Upgrade bandwidth from 4xT2 (24 Mbps) to 1xT3 (45 Mbps).
- Internet: Network traffic of 70.2 Mbps. Upgrade bandwidth from 4xT2 (24 Mbps) to 2xT3 (90 Mbps). Note: Public Internet is not a single site and can be represented by the Data Center bandwidth.

Two WAN remote sites will require network upgrades.

- Site 2_Operations: Site traffic of 11.0 Mbps. Upgrade bandwidth from 2xT1 (3 Mbps) to 4xT2 (24 Mbps).
- Site 4_Willsberg: Site traffic of 6.7 Mbps. Upgrade bandwidth from T2 (6 Mbps) to 3xT2 (18 Mbps).

Four Internet remote sites will require network upgrades.

- Site 6_Wawash: Site traffic of 5.0 Mbps. Upgrade bandwidth from T1 (1.5 Mbps) to 2xT2 (12 Mbps).
- Site 7_Jackson: Site traffic of 2.6 Mbps. Upgrade bandwidth from T1 (1.5 Mbps) to 1xT2 (6 Mbps).
- Site 8_Petersville: Site traffic of 7.6 Mbps. Upgrade bandwidth from T1 (1.5 Mbps) to 3xT2 (18 Mbps).
- Site 4_Rogerton: Site traffic of 7.6 Mbps. Upgrade bandwidth from T1 (1.5 Mbps) to 3xT2 4xT2 (24 Mbps).

Figure 9.27 **Year 2 network upgrade recommendations.**

batch process workflow platform loads analysis introduced on page 148 (Xeon E5-2667 servers were selected as host platform for virtual server deployment).

Two platform solutions will be considered for City of Rome year 2 deployment.

- High-availability virtual server configuration
- Hosting public web services on Amazon cloud

Hardware pricing will be calculated for each configuration candidate.

Year 2 high-availability virtual server configuration

The virtual server environment includes the host platform and the virtual server configuration. The host platform

City of Rome Year 2 / Types of Workflows	Software Configuration: Desktop		Web	Server	Data Source			Platform Service Times, Arc12 4 core (1 chip) Baseline				
Standard	Client	Citrix	Web	SOC	SDE	DBMS	Data Source	Client	WTS	WebIn	WebPub	DBMS
LAN_Local Clients	Client	WTS	WebIn	WebIn	Default	DBMS		Platform Service Times (seconds)				
GISDeskEditor	Default				Default	Default	SDE_DBMS	0.190				0.021
GISDeskView	Default				Default	Default	SDE_DBMS	0.096				0.011
GISDeskBA	Default				Default	Default	SDE_DBMS	0.190				0.021
WebInternal	Default		Default	Default	Default	Default	SDE_DBMS	0.036		0.104		0.014
BatchAdmin			Default	Default	Default	Default	SDE_DBMS			0.157		0.029
RemoteGISView	Default	Default			Default	Default	SDE_DBMS	0.036	0.190			0.021
WebPublic	Default		WebPub	WebPub	Default	Default	SDE_DBMS	0.036			0.060	0.007

Figure 9.28 **Rome year 2 software configuration.**

City of Rome Year 2 / Types of Workflows	User Environment: Peak Concurrent		DPM/TPM		Network	Bandwidth Mbps	NW %Cap	Platform Service Times, Arc12 4 core (1 chip) Baseline					Total baseline platform loads: Client	WTS		WebIn		WebPub		DBMS	
Standard	Users	TPH	per Client	Total	Mbps		Traffic Mbpd	Client	WTS	WebIn	WebPub	DBMS	Users	Users	Sec	Users	Sec	Users	Sec	Users	Sec
LAN_Local Clients		Services	87 Clients	LAN = 62.4 Mbps		1000	6%	Platform Service Times (seconds)													
GISDeskEditor	19		10.00	190	17.417		5.500	0.190				0.021	19.0							19.0	4.0
GISDeskView	49		10.00	490	24.500		3.000	0.096				0.011	49.0							49.0	5.4
GISDeskBA	18		10.00	180	16.500		5.500	0.190				0.021	18.0							18.0	3.8
WebInternal		8,400	6.00	140	3.967		1.700	0.036		0.104		0.014	23.3			23.3	14.6			23.3	2.0
BatchAdmin	1		249.28	249						0.157		0.029				1.0	39.1			1.0	7.2
WAN_Clients			136 Clients	WAN = 18.8 Mbps		45	42%														
Site 2_Operations			66 Clients	Trafic = 9.7 Mbps		24	40%														
RemoteGISView	66		10.00	660	7.810		0.710	0.036	0.190			0.021	66.0	66.0	125.4					66.0	13.9
WebInternal		4,000	6.00	67	1.889		1.700	0.036		0.104		0.014	11.1			11.1	6.9			11.1	0.9
Site 3_Freeberg			30 Clients	Trafic = 3.9 Mbps		12	32%														
RemoteGISView	30		10.00	300	3.550		0.710	0.036	0.190			0.021	30.0	30.0	57.0					30.0	6.3
WebInternal		700	6.00	12	0.331		1.700	0.036		0.104		0.014	1.9			1.9	1.2			1.9	0.2
Site 4_Willsberg			40 Clients	Trafic = 5.2 Mbps		12	43%														
RemoteGISView	40		10.00	400	4.733		0.710	0.036	0.190			0.021	40.0	40.0	76.0					40.0	8.4
WebInternal		1,000	6.00	17	0.472		1.700	0.036		0.104		0.014	2.8			2.8	1.7			2.8	0.2
Internet_Clients			182 Clients	Internet = 70.2 Mbps		90	78%														
Public_Web Services				Trafic = 47.2 Mbps		90	52%														
WebPublic		100,000	6.00	1,667	47.222		1.700	0.036			0.060	0.007	277.8					277.8	100.0	277.8	11.7
Site 5_Perth			2 Clients	Trafic = 0.3 Mbps		1.5	19%														
RemoteGISView	2		10.00	20	0.237		0.710	0.036	0.190			0.021	2.0	2.0	3.8					2.0	0.4
WebInternal		100	6.00	2	0.047		1.700	0.036		0.104		0.014	0.3			0.3	0.2			0.3	0.0
Site 6_Wawash			40 Clients	Trafic = 5.0 Mbps		12	41%														
RemoteGISView	40		10.00	400	4.733		0.710	0.036	0.190			0.021	40.0	40.0	76.0					40.0	8.4
WebInternal		500	6.00	8	0.236		1.700	0.036		0.104		0.014	1.4			1.4	0.9			1.4	0.1
Site 7_Jackson			20 Clients	Trafic = 2.6 Mbps		6	43%														
RemoteGISView	20		10.00	200	2.367		0.710	0.036	0.190			0.021	20.0	20.0	38.0					20.0	4.2
WebInternal		400	6.00	7	0.189		1.700	0.036		0.104		0.014	1.1			1.1	0.7			1.1	0.1
Site 8_Petersville			60 Clients	Trafic = 7.6 Mbps		18	42%														
RemoteGISView	60		10.00	600	7.100		0.710	0.036	0.190			0.021	60.0	60.0	114.0					60.0	12.6
WebInternal		1,000	6.00	17	0.472		1.700	0.036		0.104		0.014	2.8			2.8	1.7			2.8	0.2
Site 9_Rogerton			60 Clients	Trafic = 7.6 Mbps		18	42%														
RemoteGISView	60		10.00	600	7.100		0.710	0.036	0.190			0.021	60.0	60.0	114.0					60.0	12.6
WebInternal		1,000	6.00	17	0.472		1.700	0.036		0.104		0.014	2.8			2.8	1.7			2.8	0.2
Total Throughput	405	117,100											729.3	318.0	604.2	48.5	68.8	277.8	100.0	730.3	102.8

Figure 9.29 **Rome year 2: Virtual server platform loads analysis.**

determines the virtual server per-core performance. The virtual server machines are configured as 2-core servers. There is a 20 percent processing overhead for each virtual server (10 percent per core). The final server configuration performance will depend on how the virtual servers are deployed on the host platforms (for planning purposes, there should be no more virtual server core than available physical host platform core—do not over allocate).

The selected virtual server host platforms remain the same as in year 1. The virtual server platform sizing analysis and pricing are provided in figure 9.30.

Looking at figure 9.30, the baseline seconds for each tier are from the figure 9.29 workflow loads analysis. Adjusted seconds column accommodates the slower E5-2667 processor speeds. Required core calculations include virtual server 20 percent processing overhead (1.2 × 60 sec/adjusted sec). The completed analysis in figure 9.30 shows the peak virtual server machine requirements.

Platform Selection Platform Configuration	SRint2006 per Core	Total Chips
Xeon E5-2667 12 core (2 chip) 2900 MHz	41.8	2

Server Platform tier	Baseline sec	Adjusted sec	Percent Rollover	Required Core*	Core/ VM	Primary VM	Backup VM	Min RAM per VM	VM CPU Utilization
WTS: WTS Platform Tier	604.2	651.2	80%	13.0	2	9	1	19 GB	65%
WebIn: Internal Mapping Tier	68.8	74.1	80%	1.5	2	1	1	5 GB	37%
WebPub: Public Mapping Tier	100.0	107.8	80%	2.2	2	2	0	5 GB	54%
DBMS: Database Tier	102.8	110.8	80%	2.6	4	1	1	38 GB	65%
* Includes 10% processing overhead per virtual server core					Total core =	28	8		

Platform Selection Platform Configuration	Platform Core/node	Total Virtual core	Physical Nodes	Total Servers	RAM per Node	Cost per Server	Total VM	Total Cost
Xeon E5-2667 12 core (2 chip) 2900 MHz	12	36	3.00	4	128 GB	$17,185	$16,000	$84,740
Storage pricing estimates: Volume of data gigabites + 50% (data indexing) x $ per GB at required RAID level (see data storage estimates)								

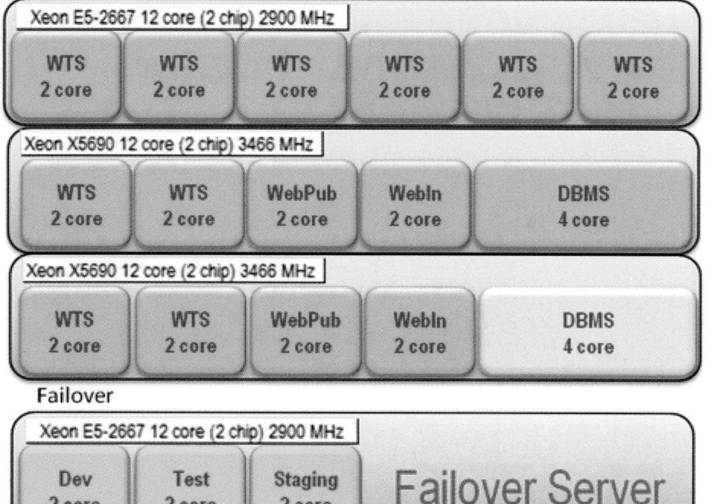

Virtual Server Pricing

Vmware	Price/ chip	MaxCore/ chip
Standard	$2,000	6
Advanced	$4,500	12

$8,000 (4 chip x $2,000)

Figure 9.30 **Rome year 2 high-availability virtual server platform selection and pricing.**

The WTS platform tier increased from three to ten 2-core virtual server nodes with projected utilization reaching 65 percent. The two Web Internal mapping 2-core virtual servers are able to handle the increased loads, with projected utilization reaching 37 percent. The two Web Public mapping 2-core virtual servers are able to handle the increased loads, with projected utilization reaching 54 percent. The database was increased to a 4-core virtual server to handle the increased loads, with projected utilization reaching 65 percent. Virtual server environment would require a total of four Xeon E5-2667 12-core (2-chip) 2900 MHz host platforms (fourth host machine was required for high-availability failover). Total cost for server platforms has increased to $84,740 (4 × $17,185 + 8 x $2,000).

Year 2 high-availability virtual server configuration with public web services deployed on Amazon cloud

The second design option will host the public web services in the Amazon cloud, reducing the number of required host platforms within the Rome data center. City of Rome worked with the Amazon cloud provider to identify pricing information for use in their design analysis. A full explanation of this platform selection and sample Amazon EC2 cost analysis is provided in appendix E.

Total Amazon EC2 cost estimate over the three year period would be $18,210. Pricing includes cost of two Amazon Machine Instances (AMIs)($12,027), data upload and S3 storage costs ($481), data out costs ($4,666), and elastic load balancing ($1,036).

Workflows	Rome City Hall Year 1				
	Peak Concurrent		Servers		
	Users	Services (TPH)	Min Physical	HA Physical	HA Virtual
GISDesk	46				
RemoteGIS	52		1 x 4core	2 x 4 core	2 x 2 core
BatchAdmin		1			
WebInternal		6,100	1 x 4 core	2 x 4 core	1 x 2 core
WebPublic		30,000	1 x 4 core	2 x 4 core	2 x 2 core
DBMS			1 x 4 core	2 x 4 core	1 x 2 core
DEV/TEST/STAGE					3 x 2 core
Spare					1x12 core
Amazon WebPublic					
Server Cost	98 Users + Services		$55,740	$111,480	$42,370

Workflows	Rome City Hall Year 2			
	Peak Concurrent		Servers	
	Users	Services (TPH)	HA Virtual	HA Virtual+AMI
GISDesk	86			
RemoteGIS	318		9 x 2 core	9 x 2 core
BatchAdmin		1		
WebInternal		16,600	1 x 2 core	1 x 2 core
WebPublic		100,000	2 x 2 core	
DBMS			1 x 4 core	1 x 4 core
DEV/TEST/STAGE			3 x 2 core	3 x 2 core
Failover Spare			1x12 core	1 x 12 core
Amazon WebPublic		100,000		2 x 2 core
Server Cost	404 Users + Services		$84,740	$81,765

Figure 9.31 **Rome hardware business case summary.**

City of Rome business case summary

A summary of the City of Rome City Hall platform configuration options is provided in figure 9.31. The results of the City of Rome year 1 analysis show over 62 percent savings when moving from a high-availability physical hardware solution to a consolidated virtual server configuration. The results for year 2 show a 4 percent overall hardware savings by moving the public web services to an Amazon cloud deployment.

Police department network bandwidth suitability analysis

The police department is located within city hall, with a separate communications network separated from the city network by a secure firewall. The user requirements summary in figure 9.17 identifies 70 GIS users located within the police department, with 15 concurrent GIS desktop editors and 9 concurrent viewers. There are a total of 20 remote police patrols that have GIS mobile devices in their police vehicles. The mobile clients connect to a mobile synchronization service that communicates real time with the remote patrol vehicles. The police department decided to use the high-availability virtual server platform configuration. The network suitability analysis, workflow software configuration, virtual server platform loads analysis, and high-availability virtual server platform selection and pricing are described in appendix E. Total cost of the police server environment is $36,770.

Other considerations

Other things to consider at this stage of GIS planning are your organization's existing policies and standards, technology life cycles, and when you're ready, writing a preliminary design document convincing enough to earn you the approval of upper management to proceed to implementation planning.

Organization policies and standards

Most GIS planners must design their organization's hardware solution and meet software and communication needs within the context of existing policies and practices. If possible, plan to comply with those policies and standards your organization may have adopted regarding system configuration. For example, you must find out if your organization has adopted an operating system standard (e.g., Microsoft Windows XP or 7, UNIX) or if they have a specific processing model (e.g., web processing), and then document—and be cognizant of—these preferences in the preliminary design report.

Many organizations have a long-established relationship with particular hardware or software vendors. These relationships developed because of the past reliability and suitability of the products and the level of service satisfaction and trust in particular vendors. Your organization may already have established maintenance agreements (common in big enterprise hardware and software deals) or have invested heavily in training for certain products. These issues cannot and should not be ignored. On the other hand, because of normal technology life cycles, no solution can ever be considered permanently set in stone. It is not uncommon for the prospect of an enterprise-wide GIS implementation to invoke major changes in technology standards across the organization. If a change in technology platforms or operating systems is required, spell out the reasons clearly. Explain how existing capabilities are critically inadequate and list the benefits of the alternatives you're suggesting.

Technology life cycles

The life cycle of technology, including the rate of technologic change, is a crucial consideration in the acquisition, purchase, or upgrade of new systems. Computer technologies are constantly evolving and becoming more cost-effective (we've all heard of Moore's law by now). Stay abreast of this and tell senior management about your strategy for keeping the GIS cost-effective. If management knows well in advance that future upgrades will be required, they will be that much more likely to approve funding for regular technology upgrades.

Figure 9.32 shows the life cycles of all of the technologies related to a major GIS implementation. It lays out how many years each technology, in turn, would be considered current, useful, obsolete, then nonfunctional, according to the following definitions. *Current* represents the period between major releases with significant functional improvement. *Useful* indicates the length of time that current software will run on this equipment. *Obsolete* shows when new releases of software will not be compatible with the equipment. *Nonfunctional* is defined as the point at which the technology is no longer worth the cost of maintenance and training. Note that networking technology and operating systems have the longest life cycles (up to three years of currency). Contrast this to workstation, desktop, laptop computers, and mobile devices, which can be expected to move from current to useful within much shorter periods.

To stay current with technology trends, attend software user conferences, read GIS publications, and establish communication with vendors and industry peers.

The preliminary design document

At the end of this stage in your GIS planning, after having determined both the data and technology requirements, you will be ready to place your findings into a report. This interim report, the preliminary design document, covers all your conceptual work so far and will mark the transition from system design to implementation planning, system procurement, and implementation. This document is a critical component of your effort: it spells out the requirements for the GIS that must be put in place to meet your organization's needs. It must win executive approval to move into implementation planning.

The conceptual system design overall—and therefore also this document—is derived from the function requirements identified in the IPDs, the data-input requirements as identified on the MIDL, and the conceptual system design for technology (addressed in this chapter). The planning process has built on itself, with the results from earlier work carried through for use in

Technology	Current	Useful	Obsolete	Nonfunctional
Network infrastructure	2–3	3–6	6–10	10+
Wide area networks*	1–2	2–5	5–7	7+
Computer				
• Server	1–2	2–4	4–6	6+
• Workstation	1–2	2–4	3–5	5+
• Laptop	1–2	2–4	3–5	5+
• Mobile devices	1–2	2–4	2–4	4+
OS software	1–3	5–6	5–6	6+
Vendor software	1–3	2–4	3–6	6+
Internet products (browsers, associated products)	1–3	2–4	3–6	6+
Data	Variable—depends on rate of decay of validity			
*Internet bandwidth increasing at 100% per year				

Figure 9.32 **2012–2013 technology life cycles estimated in years.**

later stages. At this stage, you will use these results to inform your report and back up your recommendations.

See the following focus box for details about how to write the preliminary design document. You will circulate your first draft for review and comments to the individuals involved in the GIS database design. When you are implementing a shared database model, representatives from each department intending to use the system must review your plan with a careful eye to ensure that the database concepts overall are consistent and complete.

After this comment period, you will submit the preliminary design document to the GIS committee and senior management for formal approval. Before you can move into the actual implementation planning and procurement, there must be general agreement on the overall system design, and then attention paid to the three issues we'll consider in the next chapter: benefit-cost, migration, and risk.

This book recommends presenting three reports to management:

1. **The planning proposal, to gain approval to begin the planning phase**
2. **The preliminary design document, to present the database, software, and hardware recommendations along with the IPDs and MIDL**
3. **The final report, which is presented to obtain approval for implementation and is eventually folded into the overall business plan of the organization**

FOCUS

Writing the preliminary design document

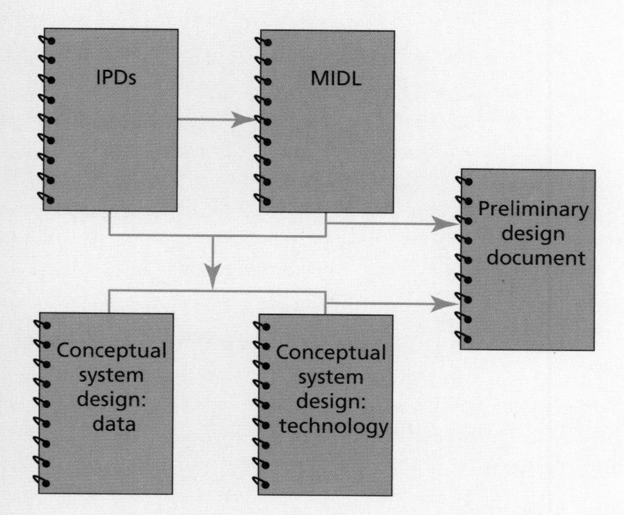

Figure 9.33
Components of the interim report to management.

The preliminary design document is the culmination of all your planning work before the end stage planning for GIS procurement and implementation. It should include the following sections.

(continued on next page)

FOCUS continued

Executive summary

The preliminary design document begins with a summary of your findings and recommendations, so it's likely you will write this first section of the report last. Senior officials, who are short on time and low on GIS knowledge, probably will read this part only, so you want it to be convincing. After you've completed the report, review it in its entirety, paying close attention to the recommendations section. Make sure the impact of the work on the management accountability framework objectives is clear. Gather this information into summary form and create the section that management will see first, the "Executive summary."

Introduction

Include a brief introduction describing the report's purpose and structure. Include a table of contents in this section.

Information products

List the information products that you propose to create, along with a brief summary for each, and point to the appendix in with the detailed information product descriptions are found. Everything depends on the information products and their contribution to the management accountability framework objectives..

Data

The data section details the design and development of the database and its effect on system design. Structure it to include the following subsections.

Dataset names

There is little to build on without first identifying the datasets. You'll find the name of each in the MIDL.

Data characteristics

Gather the data characteristics from the MIDL, and then identify the physical and spatial characteristics of each dataset. Report the medium of the source data (e.g., hard copy, zip files, DVDs, or website URL). If the data is already in digital form, identify its format (e.g., TIGER, shapefile, GDB feature class, or text files). Report the size of each dataset, map projection, scale, and datum. Carefully note any dataset that will require conversion to the chosen projection or scale. Finally, be sure to include the type and amount of error tolerance for each dataset.

Logical database model

Here you will report the database contents, including all of your data elements and their logical linkages. Begin organizing the data thematically here, clustering such topics as landownership, transportation, and environmental areas together. Include diagrams of the relationships between data elements. This step will lend some perspective on the overall database and the complexity of its data relationships.

Technology

The technology section of the document defines the remaining components of the conceptual design for your organization. Clearly identify your planned use of the technology within the following subsections.

(continued on next page)

FOCUS continued

Function utilization

Report the function utilization in the main document or as an appendix. It should contain an overview of the software functions required to create the information products. Gather this information from the steps you did earlier in summarizing and classifying your function requirements, and include your table or graph.

System interface requirements

Describe the system interface requirements you have identified. For example, your GIS applications may require a link to scanned images and documents that are already managed by an existing software program. If this is the case, specify the need to access these databases and where they reside. Include detailed information about the specific software, as an appendix if needed.

Communication requirements

In this subsection, review the existing communication infrastructure and identify the network communications required to support the GIS. Also report all locations within the organization requiring access to GIS and the number of GIS users at each location.

Hardware and software requirements

Describe the existing hardware and software standards and policies of your organization, as well as the proposed hardware and software configuration for the new GIS. If necessary, discuss any system configuration alternatives based on your organization's policies and standards. Document any issues of compatibility between the existing computing standards and the proposed system configuration. Suggest how this proposed configuration might integrate with other systems, already used or planned for use, within the organization.

Policies and standards

Report the policies or standards regarding the implementation of technology that exists within your organization. If there are no specific policies or standards that might affect the conceptual design, note this here.

Recommendations

The last section of the preliminary design document is the clear-cut recommendation for a system design. All the previous sections have been strict reportage—facts and findings from your earlier work. This section is where you and your supporting team can actually declare your advocacy for the system's final configuration.

Base your recommendations on logical decisions. Draw from the data and technology sections of this report and use your staff to help with the decisions, then make the design recommendations for both software and hardware technology. You have reached a critical point in the planning process. The assessment of needs, dta, and technology is complete. The recommendations take all of this into account and provide a basis for asking for approval to proceed with planning for procurement and implementation.

Appendixes

Consider anything supplemental but not vital to understanding the core of the design as a candidate for an appendix. For example, it's essential to provide all the IPDs and the MIDL as appendixes (this way, too, they will be at hand for reference when you write your final report, the plan for implementation presented in chapter 11).

Chapter 10

Consider benefit-cost, migration, and risk analysis

The most critical aspect of doing a realistic benefit-cost analysis is the commitment to include all the costs that will be involved. Too often managers gloss over the real costs, only to regret it later.

Rarely do the benefits from GIS become evident right away, and the initial outlay of money can be substantial. Invariably, GIS brings change—the need to migrate from old systems to new—and change can be unsettling. Also, approaching implementation, the planning team starts feeling the risk of project failure. It's important to supply yourself with information to meet these concerns. Before proceeding with planning the actual launch, you must show whether and when GIS will become cost-effective.

Benefit-cost analysis and cost models

Your first task at this stage is to do a benefit-cost analysis, with the aid of a cost model, and document both. Benefits have been estimated on an information-product-by-information-product (hence department-by-department) basis, but costs are for the GIS as a whole. The funding source may be enterprise-wide or shared between certain departments or budgets. By and large, each organization will arrive at its own formula, yet benefit-cost calculations are typically for the whole.

Benefit-cost analysis is a technique that allows you to compare the expected cost of implementing a system with the benefits it is expected to bring within a certain period of time. This comparison indicates whether or not your project is financially viable and when you can expect an actual financial return on your initial investment. A cost model charts all the costs you expect to incur during your GIS implementation, also within a specific planning period. You create the cost model during the first step of these four steps of benefit-cost analysis:

1. Identifying costs by year
2. Calculating benefits by year
3. Comparing benefits and cost
4. Calculating benefit-cost ratios

Identify costs by year

The focus of benefit-cost analysis is always within the context of a specific time frame, as is the focus of a cost model. To create a cost model for your project, set up a cost matrix that breaks down the five cost categories presented in "Focus: The cost model" by the year that the cost will be incurred. Establish columns for each year. List the data that will be available in the first year. Determine the costs associated with the hardware and software needed to produce information products as they are planned year by year. Anticipate the pace of the growth of your system—look ahead at expected demand—and plan accordingly.

What makes the model work as a true gauge of reality is your commitment to include all the costs that will be involved. Surprisingly to the uninitiated, the immediate hardware and software costs typically comprise a relatively small percentage of the overall cost model initially. Staffing and data require significant up-front expenditures. Refer to the cost model components and figure your costs for at least the first year; and then estimate

FOCUS

The cost model

Include in the cost model these four cost categories and their associated components.

Hardware and software

Estimate the yearly cost of the hardware and software you will purchase. Include the following:

- Workstations (high-end and low-end)
- Servers, including data servers, application servers, blades, terminal servers, web servers, and map servers, and cloud hosting services
- Mobile devices
- Input devices such as digitizers, scanners, GPS cards, USB thumb drives, smartphones, digital cameras
- Output devices such as laser printers and plotters
- Purchased applications and apps
- Software licenses, including extensions, extra seats, and contract services
- Maintenance costs for your hardware and software, as well as your planned upgrades
- The scheduling of incremental technology acquisition is the key to subsequent financial management

Data

As the need to generate more and more information products grows, so likely will your database. Depending on the project purpose and scope, data acquisition can easily become a significant expense, particularly over the first five years. The costs are associated with each dataset acquisition, including licensing or royalties as required by year. The costs of data conversion, development, and maintenance are included in staff time.

(continued on next page)

your costs for two-, three-, four-, and five-year windows to see what may happen over time.

As you build the cost model, the value of the planning process starts to show (or, as the case may be, its absence is felt). Since you have planned the information products and the data readiness that they depend on, you know such things as whether and when you need a budget for cloud hosting services and how big it should be. You've thought about it all, so you can trust that your expectations are realistic. Unfortunately, not all confidence is so grounded in reality. There have been well-publicized examples in years past of major GIS technology procurements that failed to specify information products up front, so they did not factor in application programming. As a result, costly delays and other unanticipated expenses ensued. The worst part of instances like these—and there continue to be too many—is that the organizations lose their chance to see what GIS can do for them. Poor planning has doomed the effort before the much-needed new business processes can even be deployed.

Calculate benefits by year

Now shift your focus to the benefits of GIS. Benefits stem from new information that allows an organization to conduct business better, faster, cheaper, or more productively. Sufficient empirical data now exists to show that you can rely on benefit analysis in GIS to be an economically rigorous reality check on your own planning. It is important to relate these benefits directly to the objectives of the management accountability framework used by the organization. This will help ensure management is invested in seeing the benefits come to fruition. Benefits should be assigned to one of the following categories:

FOCUS continued

Staffing and training

Staffing is probably the largest segment of system cost over time. Recently we have been paying more attention to staff cost in the cost model and simplifying application programming in the process.

Create your estimates for staff time by year for the following:

- Data conversion, including any necessary editing
- Data development, including updates
- Data maintenance
- Application programming and development
- Contract staff (programmer, application developers)
- System administration
- Customer support
- Staff training and travel, including the costs of travel required for training and retraining

Interfaces and communications

Finally, calculate the cost of any hardware or software interfaces you need to purchase and any communications networks necessary to support your system. Include hardwiring, hubs, switchers-routers, remote communications equipment (modems, compression DSNs), leased communication lines, software, and site preparation in your calculations.

- Savings: savings in money currently budgeted (i.e., in the current fiscal year) through the use of the new information provided by the proposed GIS (e.g., reductions in current staff time, decreases in current expenditures, increases in revenue).
- Benefits to the organization: improvements in operational efficiency and in workflow, reduction of liability, and in the effectiveness of planned expenditure (less cost, better results). In my observations of the use of GIS worldwide, this category of benefits stands out as the most significant. (See "Focus: The benefit approach" below for a useful way to quantify these benefits.)
- Future and external benefits: these are benefits that accrue to organizations other than those acquiring the GIS or outside the planning time horizon of the work. Such benefits are often underestimated in benefit-cost calculations.

The proponent of an information product should have already estimated the benefit that will result from having it. Specifically, a benefit analysis should include the following:

- The benefit category
- How the anticipated benefit will be measured
- The dollar value of the benefit measurements, within the financial context of the organization (i.e., within an existing budget category), calculated annually
- An approval sign-off by the supervisor of the proponent.

Of course, benefits will not begin to be realized immediately. Your plan predicts when each information product will be available. The question to be answered now is, when will the benefits from each accrue to the organization? One year after the first availability of the particular information product is the conservative estimate of when the benefits will be realized, even though this will depend somewhat on the nature of the benefit. It is

FOCUS

The benefit approach

When attempting to quantify the benefits of an information product to your organization, consider the following steps:

- Write sentences that describe directly what the information product will do (on the ground).
- Identify line-item budgets that will be affected by the new information.
- Identify actual change in operating procedures in subline-item budget that will result from the improved information.
- If applicable, estimate the percentage of increase in effectiveness of that expenditure as a result of the new information (quicker, more accurate, less cost, more work accomplished, better results, lower public safety risk, lower risk of damage, reduction in liability, etc.)
- As a consequence of any or all of the improvements identified above, there may be potential improvements in other activities (other line-item budget categories). Identify these activities and assess them as well. As these are *consequential* benefits, the estimates for their percentage of increase in effectiveness (in the fourth item in the list above) are likely to be modest.

desirable to establish a monitoring process to track benefit measures over time. Benefits should be reviewed at regular, preestablished intervals and coordinated with other budget reporting timelines. Costs associated with benefit monitoring should be accounted for as organizational budgetary expenditures rather than GIS costs.

Review figure 6.17 on page 69 for an example of a signed list of complete information product benefits. This is part of the IPD, revisited and refined when it's time to do the benefit-cost analysis.

Here are a few more examples of measurable benefits:

- Department of Education: Digital floor plans of schools resulted in more accurate measures of floor space. Cleaning contract ($200 million last year) renegotiated to cost $4 million less this year.
- City Planning and Housing: GIS will offer improved planning scheme information. Estimated time savings: 10 percent multiplied by 100 planners equals $600,000 annually. GIS will also improve the public inquiry service. Increased revenue from inquiries (when written responses are fee-based) and planning certificates of $100,000 annually.
- Department of Conservation and Natural Resources: Increased leasing due to improved tracking of parcels and improved calculation of lease values. Current estimates of lease values were found to be 15 percent undervalued; change in revenue +$151,200 annually.

Compare benefits and costs

After you have estimated the costs of system implementation and the benefits that will accrue from having at hand the information products created by the system, you can assess the benefits in relation to their associated costs. This assessment is the crux of the benefit-cost analysis. It will tell you if your implementation is financially feasible and, if so, when you will see a real return on your GIS investment.

You can use graphs like the ones here to visualize costs versus benefits over the life of the project. Note in figure

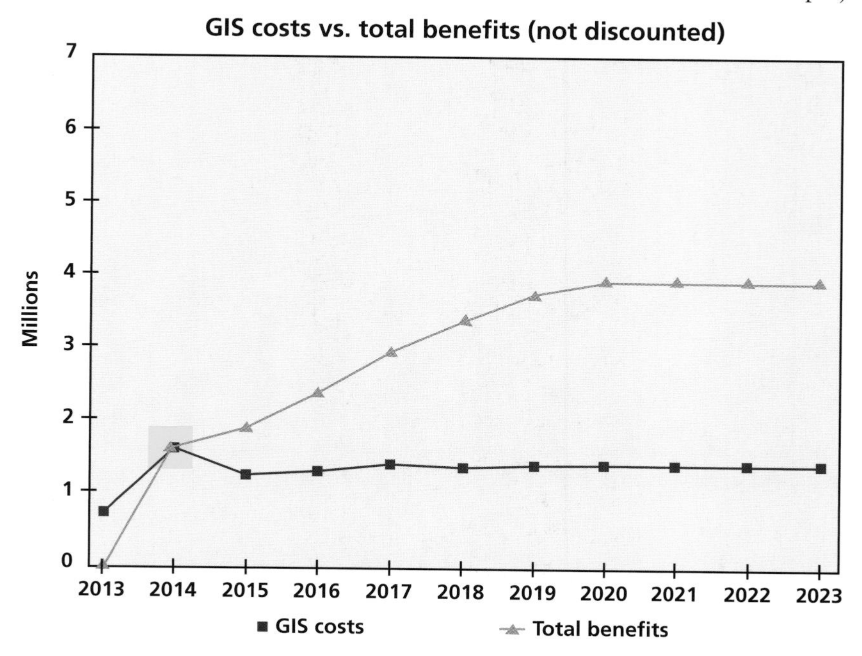

Figure 10.1 **Crossover point to a positive cash flow.**

10.1 where the two lines cross: this is the crossover point to a positive cash flow.

In figure 10.1, with values in nondiscounted dollars, costs are represented by the green line, which peaks early at approximately $1.4 million then quickly falls and plateaus. Meanwhile, the benefits (the orange line) rise steadily through the early years and then level off at a high value over time. The graph shows that implementing this GIS requires a high front-end investment, with a positive net effect only one year into the project.

You can also examine costs and benefits as a pair of cumulative numbers, side by side over time as in figure 10.2. The new benefit or cost incurred each year is added to the previous year's total to give the new number. All numbers have also been discounted (brought back) to their equivalent value for the year 2013 to remove the effects of inflation and reflect the true net present value of money.

The cumulative benefits first exceed the cumulative costs in 2015. The overall benefits of a well-planned and well-managed system typically begin to outweigh the total costs as soon as the third year. (With today's lower hardware costs, it doesn't take very long to reap the benefits of GIS.)

Calculate benefit-cost ratios

For the last step, express all costs identified in your cost model in discounted funds. In benefit-cost analysis, you use discounted costs, as shown in the bottom row of figure 10.3, in order to capture the true value of money. Discounting is most significant when the total dollar

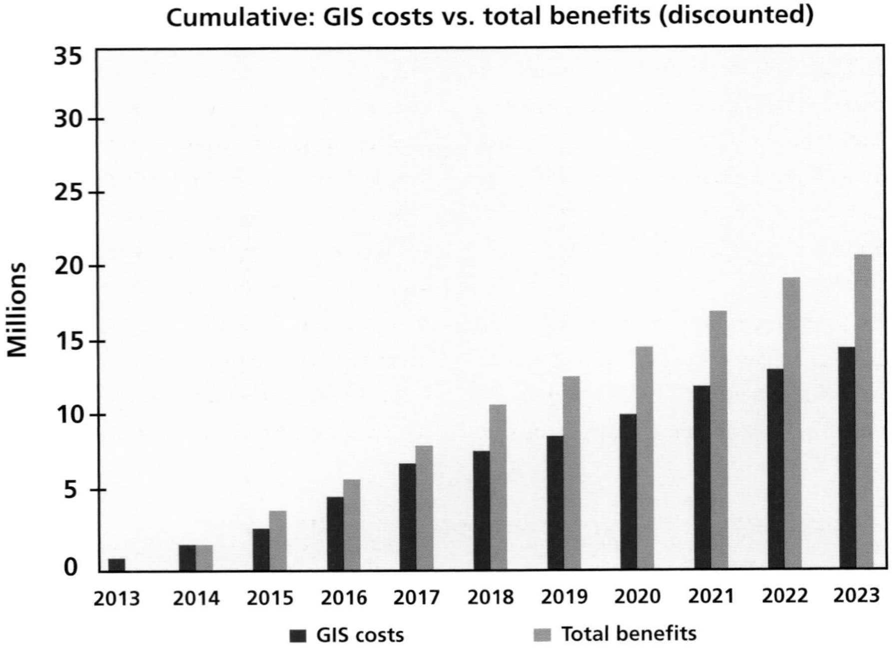

This graph illustrates the cumulative costs versus benefits with all figures brought back (discounted) to the value of the investment as if made in 2013. When the value of the investment is taken into account in this manner, the cumulative benefits exceed the cumulative costs in 2015.

Figure 10.2 **Cumulative benefits exceed cumulative costs.**

value is high or the cost of money is high (high inflation rate). Costs that spread over a period of years into the future must be discounted to the base year using a discount rate to remove the effects of inflation.

The base year is the current year or the year in which the benefit-cost analysis begins, whichever is the earliest. For example, suppose the base year is 2013 and the cost of a maintenance contract is $100 per year. Because of inflation, a cost of $100 incurred in the year 2014, is less in real terms than a cost of $100 incurred in 2013. Similarly, if $100 of benefit accrues in the year 2015, this is worth less in real terms than a benefit of $100 realized in 2014. To assess investment decisions properly, the benefits as well as costs must be discounted to a common year—the base year—to remove inflation's effects.

As an example, using 4 percent as the current discount rate would mean that $100 last year is worth $96 this year. Check the discount rate being used in your organization. Identify the benefits over time and give these values discounted to the base year, as you did for costs. Calculate the net present value and benefit-to-cost ratio for each year of your analysis. The net present value (NPV) is calculated by subtracting the present value of costs (PVC) from the present value of benefits (PVB):

NPV = PVB - PVC

A positive NPV is the usual decision criteria applied for acceptable projects. The benefit-to-cost ratio (benefit: cost) is a fraction that represents the return (or benefit) on the investment (cost). Usually, the denominator (the cost) is reduced to 1:

B:C = PVB/PVC

For example, if a $100,000 investment yields a return of $260,000, the benefit-to-cost ratio would be 260,000:100,000 or 2.6:1. Investments that have a ratio greater than 1:1 will at least break even.

You can assess how changes in key parameters might affect the overall results. To do this, you need to rework the calculations for different implementation scenarios. These scenarios might incur different costs or lead to different benefits that must be accounted for in the analysis. This is called *sensitivity analysis.*

These basic steps have been adapted from those presented in Tomlinson and Smith (1991). You can find more detailed information on their published methods for conducting GIS benefit-cost analysis by referring to the *International Journal of Geographical Information Systems* 6(3): 247–56.

Migration strategy

Crafting a good migration strategy is dependent on the scope of the effort. Obviously, the installation of a small system in one department requires a radically different strategy than a major "roll-out" of an enterprise-wide system in a large organization. In almost all cases, however, the new GIS will be implemented in the context of existing data-handling systems that need to migrate from the old to the new. So regardless of the scope, you need to plan the sequence of events in this process and

Cost per year	2013 (base year)	2014	2015	2016	2017	2018	2019	2020	2021	2022	2023
Actual cost of maintenance contract	$100	$100	$100	$100	$100	$100	$100	$100	$100	$100	$100
Cost discounted to base year (using discounting rate of 4%)	$100	$96	$92	$88	$84	$80	$76	$72	$68	$64	$60

Figure 10.3 **True value of money used in the cost model.**

provide timelines for dealing with the old systems as well as the new.

Legacy systems and models

The term "legacy systems" is an IT euphemism for an existing technology platform that will be phased out by new technology. But legacy systems also represent the current way of doing things, and these things—in the form of the business processes used to carry them out—must also be migrated. Complicating matters, because they support ongoing operations, these systems and processes must be migrated seamlessly into the new system with minimum disruption to business. This is quite a tightrope to walk.

Just as you have allowed time in your planning for new system acquisition, training, and break-in, so must you allow time for phasing out legacy systems and reallocating resources. This usually means a planned period of overlap between legacy systems and the new system. The annals of IT professionals are littered with stories of organizations that terminated critical legacy systems prematurely—on day one of new system implementation—with disastrous results. The key to not joining their ranks is to resist pulling the plug on any legacy system until you have proof that all of the needed data and functionality has been reliably migrated to the new platform.

The migration of existing business models can be a real challenge. Many have taken years to develop and refine, and they support business processes that must carry on within the organization. Since the need for the business model or process continues, there are three options for migration:

1. Rebuild the model so that it can be implemented in its present form into the new GIS-centric technology.
2. Improve the model, perhaps by replacing some of the assumptions with actual measurements using the GIS.
3. Abandon the model to design and build an entirely new model that takes advantage of GIS capability and makes new workflows possible.

Some of the questions you will ask yourself include the following:

- How difficult is it to map the process or model into the GIS data structures?
- How long will it take to do this?
- Who will do it? Are the people who designed the original model still available?
- What are the new GIS capabilities that you would like to access—can they be integrated into the old model or is a model redesign and rebuild necessary?

In considering these issues, you will identify many of the impediments to and possibilities for an appropriate migration strategy.

The migration from legacy systems to new GIS architectures is extremely dependent on the vendor offerings, both those already in place and the new ones to be acquired. For example, an Internet map service was developed four or five years ago from an older geographic (coverage) file system. A significant amount of software was written to create a publication database from the older file system.

Let's say that today your organization is considering replacing this old application with newer, server-based technologies. In addition to the cost of acquiring new software, it is also time to upgrade hardware and migrate the entire database to the new database platform. The most significant benefit of moving to the new technology is that publishing is accomplished "out of the box" and there is virtually no new application software to write. You fully expect that the new service will be more reliable than the old application, and that the cost of maintaining the old application code base will be largely eliminated, because with the hardware getting old and reliability expected to decline, keeping the old application would not be worth the associated costs.

New considerations

Moving from old to new systems or applications brings up new matters to consider. Present management with those that require attention:

- Age: Is the system or application slowing the forward progress of the organization?
- Costs: Consider the costs of a new system, developing a new application, and transition. Will capable staff be available? Will training or hiring be required?
- Benefits: Is it necessary to do this transition now? Weigh the benefits to be gained by the transition.
- Future transitions: Will the proposed transition meet future business needs? Will legacy applications work alongside new applications developed on the new GIS in the future?

There are also technical considerations. Again, it is easy to find oneself extremely dependent on vendor offerings—those in place and new ones acquired. The movement to enterprise operations also involves dealing with data standards and data integration issues related to mapping from one schema to another. With *gap analysis*, you need to identify any functions not yet available in the new GIS but possibly required by one of your information products. If custom applications need to be built for them, prepare for the wait time by managing expectations.

Pilot projects

Some organizations are more comfortable taking the GIS plan and implementing it incrementally via pilot projects. A pilot project is essentially a test run for part of—or for a small-scale version of—your planned GIS. For example, an organization with offices in multiple regions may want to establish the pilot GIS first at headquarters and in one of the regions.

This way, before wider implementation, users can build experience and gain an understanding of the kind of administrative and communications problems that might be encountered. Similarly, within one department it may be wise to focus on a selected subset of information products or even subsets of the database as first steps in implementation.

Pilot projects are useful in demonstrating the planned GIS to management and potential users. They also serve in evaluating the performance of the proposed system, solving data problems before the final cost model is developed, and verifying costs and benefits.

State your intentions clearly, though, and be aware that pilot projects are often conducted with ulterior motives. Vendors tend to encourage pilot projects on the assumption that once an organization brings in their system, it is more likely to buy it. Indeed, pilot projects can bias the procurement process. Be wary of the ploy of some senior managers who use a pilot project as a way to get GIS proponents off their backs without spending very much money. People in lower management lean toward pilot projects sometimes simply because they can get the GIS moving without the "hassle" of a thorough planning process. Clearly stating the objectives will help you avoid these pitfalls of the piloting process.

Avoid pilot projects if any of these are the motivating factors. An appropriate pilot project should always be launched in the context of your planning and should, in fact, be part of the plan from the beginning. If you have planned adequately and now want to start implementing incrementally, a pilot project may be a useful approach.

Risk analysis

To ensure a successful GIS implementation, you must thoroughly evaluate the risks associated with your implementation strategy and the potential for project failure. The basic approach to risk analysis is to consider the questions that emerge within each of the five steps that follow:

1. Identify the types of risk involved in your project.
2. Discuss the nature of the risks in the context of the planned implementation.

3. Describe the mitigating factors that will minimize the risks.
4. Assess the likelihood and seriousness of each risk to the project and give it a score.
5. Summarize the level of risk. (Summarize the scores from step 4 into one final score with which upper management can evaluate the project in terms of overall acceptability of the risks involved.)

Identify the risks

To identify the types the risks involved with your project, evaluate the following factors:

Technology

- Is the technology being adopted new?
- Is this the first release of the software or hardware? If so, does it have bugs or flaws?
- Are there gaps in the technology that prevent it from fully supporting your needs? If so, do you need to enter into a contract with the vendor before acquiring the system to ensure that the gaps are filled?

If the proven technology cannot create more than 80 percent of your information products, you are in a very high-risk category.

Organizational functions

- Can you foresee any functional changes in the mission of the departments or changes in departmental functions or workflow?

These changes can take a long time and add complexity to your project, which increases risk.

Organizational interactions

- Organizationally, are multiple agencies involved?
- Are they geographically dispersed? Working with multiple agencies or multiple locations adds complexity—and its incumbent risk—to your implementation.
- Are changes in management required?

These changes may take a long time; you need to determine if you can succeed without change or during a protracted change process.

Constraints

- Are there budget constraints?
- What is the timing of the project?

Adequate funding and a realistic time frame are essential for success.

Stakeholders

- Are there stakeholders at multiple levels—federal, state, local, and private sector?
- Is involvement from the public, the media, and lobby groups required?

It is important to involve all stakeholders in the process so they buy in to your solution; yet the greater the number of stakeholders, the higher the risk. Negotiating and coming to agreement during the planning process can mitigate these risks.

Overall complexity

- What is the overall complexity of the project?
- Are there federal or other regulations you must meet?
- Are multiple vendors involved?

Complexity in your implementation increases the amount of time you will need to deal thoroughly with each issue.

Project planning

- Is your project planning well defined?
- Is your implementation strategy consistent with the existing business strategy?

Realistic planning minimizes risk. If the objectives of the project are not well defined, you may spend large amounts of time and money on the wrong things.

Project management

- Are you using proven methods?
- Is there built-in accountability?
- Is there built-in quality control?

Project scheduling

- Are scheduling deadlines reasonable?
- Do you have project-management tools to identify project milestones?

Project management and project scheduling are necessary to keep your project on time and within budget.

Project resources

- Do you have adequately trained staff?
- Is there a knowledge gap in your organization (see "The knowledge gap" on page 179)?

If you do not have adequately trained staff, you will need to develop a plan to acquire or train them.

Discuss the risks in context

After identifying the risks, take each one and talk it over in the context of the planned implementation. For example, can your staff handle the new technology, or do you risk creating (or widening) a knowledge gap with this implementation? Discuss the new technology in relation to your staff's existing skill level until you come to a better understanding of the level of risk faced and how the risk can be mitigated.

Describe ways to mitigate the risks

Once each risk has been identified and discussed, consider mitigating factors and methods that could minimize the risks. Regarding the knowledge gap, two provisions could be made to reduce the risk of implementing new technologies with an untrained staff:

- Assess the current staff's skill level with an eye toward developing a training program.
- Decide to purchase software only from companies that have an established training program and budget for the necessary training.

Assess and score each risk

Assess the likelihood and seriousness of each risk to your project and give it a score. To quantify the risk associated with a knowledge gap, you could ask the following questions:

- How likely is it that your staff's skills will be insufficient to work with the new system?
- How seriously will this affect both project implementation and the agency or department itself?

If you have addressed your staff's skill level, you can assess the likelihood of the risk. You can determine the seriousness by relating the risk to the effect it will have on your system implementation and organizational functionality. Assign each risk factor a numerical weight that reflects its high, medium, or low level.

Summarize the level of risk

The final step is to summarize the risk factors. From the sum total of the scores produced in the previous section, compute the average. Use this average score to determine if the project's risk level is acceptable to the organization. It is the organization taking the risk, so don't be surprised if it has its own approach to risk analysis. Major organizations often do. If yours does, follow its methodology for assessing risk. If the planned project meets its standards for level of risk, it's time to plan your GIS implementation strategy.

Chapter 11

Plan the implementation

The implementation plan should illuminate the road to GIS success.

You know what your organization needs; it's time to ask how to get it. What is the timeline going to be? Are there obstacles looming to stop us from implementing GIS? Do we need to add staff? The planning methodology's last stage begins with considering issues that affect GIS implementation and ends with presenting your implementation plan to management.

After a few last tasks and addressing all the issues left to consider, you will spell out your recommendations in the executive summary section of your final report and in a professional presentation to the executive board. Your report should do several things very effectively:

- Recommend an actual strategy that outlines the specific actions required to implement your GIS
- Highlight the latest (if any) extra implementation actions or special concerns that have been identified
- Include a list of actions and concerns for senior management to consider
- Reflect all of the previous planning work and include all of the relevant documents to justify your recommendations

The supporting documents may be stored separately as text files, PDFs, or scanned documents as appendixes to the final report. The report itself begins with the executive summary section, followed by six sections (described on pages 196–197):

1. Strategic business plan considerations
2. Information product overview
3. Conceptual system design
4. Recommendations
5. Timing
6. Funding alternatives

After submitting it to the management committee for review, then making the necessary adjustments, you can make your presentation to the executive board.

This presentation is important on several levels, but don't expect it to be the silver bullet. The executives should agree with your implementation plan already because you have informed them by keeping them in the loop all along. If you haven't and they don't, no presentation is going to convince them. On the other hand, these people have backed you up for a year or more, paying for your every move, so they deserve a presentation of the highest quality to tell them what you've been doing. If you have done your job, they probably will have read every word of your report beforehand and are expecting this presentation to be the opportunity to nod their approval. So lay before them what there is to approve, specifics that spell out such things as whether their GIS should be a cloud-based system or not; what budget year the costs will fall into; how many extra staff positions are needed, to be paid from which budget; and so forth.

Anticipate the challenges of implementation

Implementation planning is a critical juncture in your efforts because this is when the impact of the underlying change (which can be profound) will be felt. Implementation is when the messiness of the real world intrudes on your so-far orderly planning process. Only through a clear-cut path toward final implementation can we expect our efforts to result in positive change.

Since developing an implementation strategy involves many simultaneous tasks, teamwork comes into play. You will form a GIS implementation team, if you haven't already. You will contact all the organizations and departments involved, if you haven't done so already, in order to synchronize your efforts and troubleshoot any problems. To facilitate technology procurement later, early on you will invite vendors to submit proposals as to how they would deploy hardware and software for GIS most cost-effectively.

Meanwhile, you will consider all the various strategic components that enter into your GIS implementation, including the all-important staffing and training requirements, the timeline, allocation of tasks, costs, and your recommendations for alternative strategies and for mitigating risk. You will draw on information about data and conceptual design you've already gathered and referenced in your preliminary design document to make specific recommendations.

This chapter addresses the tasks and issues involved in planning for implementation. Your plan's goal is to win funding for GIS implementation. Yet, this measured, step-by-step approach serves another purpose: by

FOCUS

Key issues to address in implementation planning

- Department staffing
- Training regime
- Funding
- Institutional interaction requirements
- System requirements and data sharing
- Legal review
- Security issues
- Existing computing environment
- Migration strategy
- Risk analysis
- Alternative implementation strategies
- Project timeline
- System procurement

facing—before actual implementation—the issues you must consider in order to compile the plan, you further buffer your project against failure. Ensuring a successful implementation the first time out can be vital because you won't always get another chance.

You need support from users almost as much as you need funding. Keep in mind, with the introduction of any new information system into an existing operation, there are bound to be bumps in the road. Even if you adopt industry-standard hardware platforms and use only well-tested GIS software, your application programs may still have bugs during their first releases. GIS managers oversee three interconnected items: budget, schedule, and functionality, any one of which affects the other two. You should prepare users for this, to manage expectations, while at the same time letting them know that their clearly communicated feedback will help you fine-tune the system into well-oiled machinery.

In one sense, this period before moving into actual implementation is the last step in the planning process. Obviously, the planning for GIS continues as long as there is a system to operate, but that original window of planning before a project launches will soon be shut. From then on, everything will happen in real-time in a production environment. For your implementation plan to become a successful guide, you must devise it by thoroughly considering all the factors that apply to your situation. Addressing the issues described in this chapter should prepare you well for making your GIS implementation strategy recommendations.

Staffing and training

No part of GIS planning is more important than staffing and training. It would be almost impossible to overstate the degree to which a successful GIS is dependent on the staff that builds it, manages its evolution, and maintains it over time. The best GIS plan in the world will not launch a successful GIS—that takes people.

Discuss staffing issues with each department affected by GIS implementation. Staffing a GIS is a long-term operational cost and a major expense for all systems.

The first priority is to establish the leadership teams who will guide the GIS implementation. (See "Focus: GIS leadership teams" on page 176.) While the implementation team may do all of the work in a small organization, most organizations will need additional staff. The number of staff associated with a GIS depends on the nature and size of an organization. Large organizations, with widespread user groups and complicated applications, could require a wide range of staff, from network and database administrators to GIS specialists and programmers, and of course managers to oversee the work of these individuals. A distributed staffing model for multisite enterprise projects calls for site facilitators who oversee the GIS implementation and operations at each location, reporting to and collaborating with a head GIS manager to ensure site operations are in line with the overall project objectives. In small organizations, staff members may have to wear many or all of the hats, but that can also work. For example, you may be the sole GIS manager of a skeleton team, in which case one or two people may have to take charge of the GIS planning, system design, data acquisition, and system administration. A lone GIS analyst may have responsibilities spanning different software packages and applications.

The basic staff positions required for a GIS implementation all have associated skill requirements. (These are described fully in appendix A, along with much more on staff requirements and training options for bringing both GIS staff and end users up to speed.)

Whenever GIS remains underused, a primary cause is lack of staff training. If people cannot make the system do what they need, they'll quickly abandon it and stick with the old, safe way of doing things. And who could blame them? They have a job to do, and the mere existence of a high-tech GIS system is not going to help them. It's not going to run itself either. It takes a live, thinking human being to frame a spatial problem in the context of a GIS.

FOCUS

GIS leadership teams: Establishing responsibilities

Before starting a GIS implementation, it is important to consider how the most productive relationship between management and GIS staff could be organized. Clarification of roles and responsibilities and establishing good lines of communication will greatly aid the implementation of any GIS. In the model proposed here, there are two leadership teams, each with a different role in the implementation phase. This paradigm provides structure and a system of checks and balances, but keep in mind, communication between the groups is essential. The GIS manager is the constant and should be considered the liaison between the teams. Again, these are only leadership teams; they by no means describe the extent of the GIS staff needed for a project.

As discussed in chapter 4, a *GIS management committee* helps you make decisions regarding project management, changes in project scope, and what to do if your project gets behind schedule. As the fundamental link to senior executives in your organization, the management committee should take an active role in the GIS decision-making process, helping you decide, among other things, when you need to reprioritize or scale-down a new application or project extension (see figure 11.1). Although the management committee should be formed at the beginning of the planning phase, its work is vital throughout implementation and the entire life of the GIS project.

Responsibility	Description
Review the implementation team reports.	GIS management committees should meet regularly to review reports from the implementation team. It is important that the management committee be actively involved in the decision-making process.
Review GIS project status.	During the management committee meetings, the current status of the GIS project relative to the implementation milestones should be reviewed.
Identify and give early warning of problems.	Reviewing the progress of the project compared to its schedule will help to identify problems as early as possible and allow adjustments to be made.
Make necessary adjustments.	Every time a GIS project requirement changes or something needs to be added or deleted, the change should be documented. If you can manage a project within the scope that was already identified, the likelihood of finishing it on time is good. One way to manage the scope is to put the management committee in the position of determining when you will expand a project and when you will not.

Figure 11.1 **GIS management committee responsibilities.**

The makeup of a management committee will vary according to an organization's nature and size. For enterprise-wide systems, it is strongly recommended that the CEO of the organization be on the management committee. The GIS manager should also be included. (The GIS manager is part of every team; this ensures a shared vision and clear communication.) The heads of finance, administration, and information should also be on the team (figure 4.4 on page 23 shows the management committee team members).

(continued on next page)

FOCUS continued

The GIS *implementation team*, which may include some of the same members who were on the *planning team* (also shown on figure 4.4), oversees design implementation, reviews progress, and identifies problems as they arise. The team reports to the management committee. The team can be as small as one or two people at a small site or an initial four or five people at a medium site (it will tend to grow). For most organizations, the implementation team should include the GIS manager, at least two business representatives (in-house end users of information products, or someone who represents the needs of external clients such as a marketing expert or public liaison), a GIS data specialist, a GIS technical expert to reinforce the technical perspective of the GIS manager, and a senior GIS software engineer who will head the system development staff (see figure 11.2). At large enterprise sites, it may be necessary to assign site facilitators to each location, who will oversee separate planning and implementation teams at each site. If this is the case, it's important that all site facilitators work closely with the GIS manager or GIO to ensure project alignment. (In fact, during the planning phase the site facilitators should report directly to the GIS manager, who will act as project manager during the planning phase. Once the project has been approved and implementation is underway, however, the GIS plan is made part of the overall business plan, therefore the site facilitators would again report to their department supervisor. This temporary change to the organizational hierarchy won't make everyone happy, but it is a proven method for successful enterprise planning.)

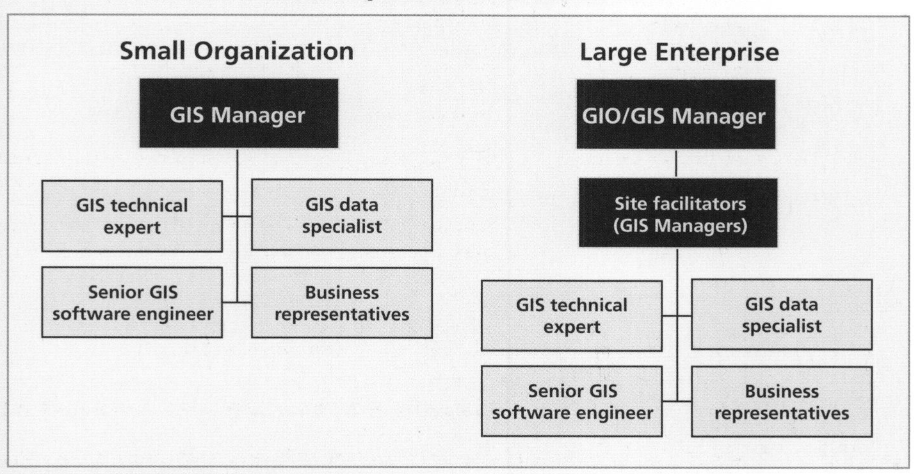

Figure 11.2 **Recommended members of the GIS implementation team.**

(continued on next page)

FOCUS continued

Whatever its size, the implementation team's tasks are consistent; their primary functions are shown in figure 11.3.

The implementation team manager is the GIS manager or GIO, whose roles and responsibilities evolve once planning ends and implementation is launched. Ensuring proper communication between staff and management continues to be a priority, as sustained buy-in at all levels will facilitate successful implementation. He or she must also keep track of the management accountability framework objectives that will be affected by GIS implementation—are project objectives to this end being completed on time and on budget? Team members must be managed, as must user expectations. The GIS manager must anticipate, share, and correct problems that will inevitably arise during implementation. The GIS manager position is crucial indeed, perhaps more in the post-planning stage. You can expect implementation to be somewhat stressful, but the fruits of labor will be plentiful.

Below is a time management recommendation for successful GIS managers:

- 30 percent strategic change
  - be part of the solution to critical problems
  - create a viable technology infrastructure
- 60 percent maintenance and operations
  - quality assurance
  - employee development
- 10 percent outreach—staying viable
  - Help others be successful
  - External involvement

Responsibility	Description
Solve design problems.	The implementation team discusses and resolves design problems as they arise.
Focus activities on the critical path.	The critical-path method is one of a number of project-management tools that can be used. It identifies the tasks and the sequence of tasks most critical to complete the project on schedule. Noncritical tasks are also identified. These are tasks that can be delayed without affecting the overall project schedule. The critical-path method is often useful for focusing the implementation team on the most important activities in the development of the project. Team members should be aware of "scope creep" and rein it in before it gets too far.
Balance workload assignments.	The implementation team is responsible for balancing the workload and assignments given to staff.
Meet regularly and report progress.	The implementation team prepares periodic progress reports (usually monthly) for the management committee. The implementation team should have regular development meetings. Initially, these might be daily meetings, then weekly. You should set up a formal schedule and reporting process where every week the team meets to at least touch base. Too often, a software expert will make some assumptions about requirements and then spend one or two weeks creating something that does not work. Frequent meetings are a way of trapping problems such as this and getting back on track quickly.

Figure 11.3 **GIS implementation team responsibilities.**

In order to get meaningful answers to their questions, people need to know how to apply the tools to the work as they understand it. Train, train, and then train some more will become your rallying cry.

The knowledge gap

Often organizations must revisit the areas of staffing and training in response to changes in technology. If their current staff lacks the skill sets to make the best of technologic advancements, some combination of new hires and training of current staff can fill the gap. The term *knowledge gap* describes a phenomenon that results from the new capabilities of technology growing faster than an organization's ability to use them. In the case of GIS technologies, there was no knowledge gap before 1990: people were still more capable than their GIS systems (figure 11.4).

By 2000, the rate of GIS development had risen above the normal growth trend of institutional management skills. This means that systems are now more capable than people, and the ordinary incremental growth rate in skills within an organization does not keep up with developments in technology. Recently, the relative curve of GIS development has leveled off somewhat, but management still has a lot of institutional learning to do before truly making use of the full capabilities of GIS. The projected knowledge gap has clear implications within an organization's training strategy and budget commitments, but it also bears on the ability of higher learning institutions to provide some of the needed skills training.

The degree to which an organization can improve its collective skills base will have a direct bearing on how well and how far the technology will be adopted within the organization. These skills are the tools that an organization uses to actually leverage technology in order to increase productivity. Only by recognizing that a knowledge gap exists and implementing efforts to fill it in can one realize that the knowledge gap is also an opportunity gap.

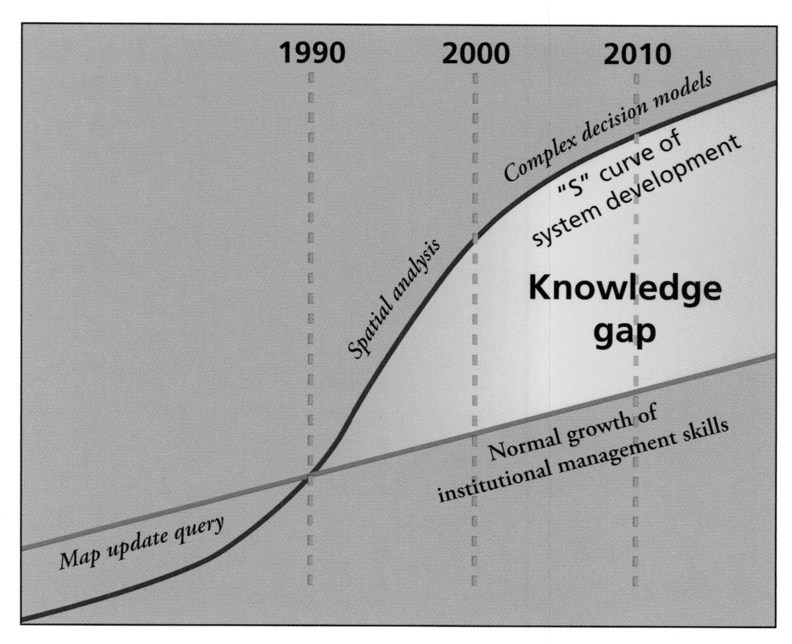

Figure 11.4 **GIS technology: Accelerating faster than the skills to use it.**

Some organizations attempt to circumvent the knowledge gap by hiring consultants to operate the GIS, under the assumption that these "hired guns" will bring the needed knowledge into the company. While this is certainly a possibility, there are few instances where it can be fully recommended. Hiring a consultant is just another way of buying the skills that you really need in your organization. It is a short-term solution to what may be a long-term requirement.

Consider the tale of a Canadian oil company that spent $200,000 implementing a complex environmental GIS application. When the application was finished, the two key staff members left, and there was no one remaining at the company who could use the system. Everyone involved in the planning and implementation must have felt like they'd been wasting their time. The moral of this story is that the host organization needs smart people in place who can really use the system.

Training programs

After determining the staffing needs for your GIS implementation, assess the GIS capabilities of your existing staff and determine an appropriate training program by personnel category. Emphasize that continuous employee training will be necessary for all GIS staff and spell out how it should be included in the budget. The all-important core GIS staff will require ongoing training to keep them current on new methods and technology. Recommend an employee training program that provides for the necessary levels of GIS staff.

Building a strong internal staff is a process, not an event, in most cases; the process itself can foster a dynamic atmosphere of opportunity. Highly selective hiring practices and ongoing training are a must, of course. But you must also create conditions that are favorable to learning and foster independent thought. This is also part of the forward-thinking approach to managing change. Provide interesting and challenging tasks, a supportive management environment, and continuing opportunities for knowledge development (including training and formal interaction with other GIS professionals).

Training delays for key GIS personnel or lack of early mentoring for site facilitators can impede the progress of the project; in order to uphold timing commitments, be sure to make training a critical priority. Furthermore, change in organization (particularly in enterprise site managers) can dramatically slow down progress, even to the point of halting implementation altogether. To avoid this, formal legacy education must be part of the personnel qualifications before and during the implementation phase. GIS project briefing notes should be compiled throughout the entire process; these should be required reading for all new key personnel.[1]

Use appendix A on staffing and training as a guide for the required training and its delivery.

GIS funding

Funding needs should be anticipated at the outset of planning, but it's not unusual for costs to rise, or for funding to be pulled back due to organizational budget cuts. In such cases, adequate funding for GIS implementation may not materialize because a senior manager is not motivated to find the funding among other funding priorities. This situation is the management committee's responsibility to anticipate, discuss, and resolve. To inspire GIS managers or site facilitators to effectively communicate the benefits of GIS (and thereby retain needed funds), a bonus plan might be considered. Bonuses are directly connected to tasks in the organization's strategic business plan and/or management accountability framework (in

1 The briefing notes should cover all stages of the planning process. It is especially important for new senior managers to be well-briefed in the process. Do not forget to include the benefit-cost analysis as it relates to the management accountability framework of the organization, as not everyone will be trained in the concepts of benefit analysis, and the measures of the benefits gained from GIS usage will directly affect budgetary decisions and may even impact performance bonuses.

which, once the project plan is approved, the timely production of information products should be incorporated.

Some aspects of GIS implementation should be regarded as necessary ongoing commitments and hence be suitable for A-level funding. These include the work involved in GIS data management and the staffing commitments for the GIS manager and his or her key staff.

Organizational issues

Management and organizational issues come into play at all stages of the planning process; at the time of GIS implementation they take precedence. Implementation becomes more complicated when working with other organizations, but it's probably worth it. The richer database and more utilitarian system that ensues will actually get used, allowing for further GIS applications to be created.

To succeed in your GIS efforts you must involve all the stakeholders in the planning and implementation process. A stakeholder is any person, group, or organization with a vested interest in your project now or in the future. These stakeholders could include organizations with which you have interactions, as well as the end users of your planned GIS-related services (the public citizen or the business customer), the media, and lobby or special-interest groups.

You may have already made contact with other organizations and partners; now you must revisit and clarify these relationships and firm them up where necessary.

These relationships will all be different. Some may be more formal and even include legal dimensions. For example, a city planning department might be legally required to immediately report the opening of new streets to the keepers of an emergency route mapping application. Less formal relationships might evolve with other organizations sharing common causes or purposes, such as a network of conservation groups agreeing to share environmental data for their mutual benefit.

You must consider how the relationships that your organization has with these disparate external partners will affect your GIS project. Ask yourself the following questions:

- Who among these partners might hinder or prevent your system implementation (i.e., are there any showstoppers)?
- Who needs to be kept informed about your project to ensure that you have continued support?
- What will happen if another organization fails to maintain its commitments to your GIS project?
- Who is responsible for managing the relationship?
- What would happen if the relationship ended?

Institutional interaction

If there is a need to interact or cooperate with other organizations, agencies, or departments in the course of your GIS project, consider whether agreements relating to the responsibilities of those involved are necessary. Such agreements should, at a minimum, be documented in a memorandum of understanding (MOU). If the system depends on these interactions, formal contracts are required.

Your preliminary design document lists the work to be done by the system, including the information products that have to be created, the system functions required, and the data needed in the database. Before recommending the actions required to put the system in place, be sure you've taken into account data requirements of all the departments and other organizations sharing data within your GIS. Enterprise-wide and community (or federated) GIS systems are starting to proliferate, so the mandate to include the data requirements of every organization and department involved is worth repeating: all should be included in the planning process.

Data-sharing arrangements

Increasingly, your GIS will require data from several sources, which leads to the need for a systematic data-sharing relationship with one or more other departments or organizations. Your implementation strategy must acknowledge the nature of any data-sharing arrangements that will be part of the GIS. Armed with your clear understanding of the information products to be created and the data requirements of your GIS, contact each of the agencies, organizations, or departments that will be involved. For example, in the case of state or municipal organizations, these contacts might include local government organizations, state organizations, federal organizations, partnerships, and multiagency bodies.

If your plan calls for data fundamental to your GIS to come from outside your own organization, it is sometimes necessary to draft a formal agreement such as an MOU or a legal contract, especially if the data is sensitive in nature. This agreement should address the following:

- What will happen if the other organization fails to supply its portion of the data?
- Who is responsible for correcting and updating data in a timely manner?
- What backup and security is in place to ensure continued access to the data?
- Who will decide what further data is collected and shared?
- Who will fund the data gathering (including maintenance and updates)?
- Who will be responsible for coordinating the regular data administrative tasks?
- Can the data be shared with the public?

Since data exchange can reduce the costs of data acquisition (typically a significant cost of a GIS implementation), there is a strong spirit of data sharing, particularly among US federal and state agencies. Although that has been tempered somewhat in this era of intensified security concerns, sharing can be beneficial.

Oftentimes before or during implementation, organizations discover that other organizations have data previously unknown to them or information that they consider of value. For example, in Australia, six or seven state government agencies were found to have data potentially useful to one another. While there had been little or no sharing of data before, during the GIS planning process they recognized that simply by sharing the data they had already developed, they could each produce new and better information products with significant benefits to their agencies. In other words, sometimes creating interagency relationships can yield substantial financial benefits for all parties. It's becoming more common for many organizations to partner in creating spatial data frameworks, deliberately shared systems of data packages, models, maps, and services (typically a cloud-based configuration) containing the best available information using common specifications and standards. Creating a spatial data framework requires clear communication and partnership between agencies in order to come to agreement regarding metadata expectations, where and how data is stored, data quality, desired imagery resolution, which data model to use, naming conventions, and a data distribution policy, for starters. Beyond sharing data, multiple agencies may share GIS applications as well, embracing the community GIS model.

Important technical issues related to data sharing can trip you up if ignored, including the data format and accuracy, the metadata standards, and the data's physical location. Two key questions should be answered to identify how multiagency or multiorganizational data sharing affects your project:

1. Which information products require multiagency data?
2. Is the benefit that will accrue to the agencies from the new information products the reason for multiagency planning?

If the answer to the second question is yes, you have a strong reason to cooperate on data sharing (or even application development), and you might also have the leverage to request joint funding of the planning process.

Once the data is developed, it is important that each organization and department be able to use it properly, so another task on your list may be creating clear user guides and metadata.

Since data can be a large cost of your GIS, sharing data may make sense for all concerned; that doesn't mean it's always easy. Particularly in large-scale enterprise or federated implementations, you may encounter resistance to data sharing between departments or agencies for several reasons:

- Worry that the data may be poor and reflect badly on the department.
- To some, data means power; possessiveness may rear its head ("it's mine").
- When data represents responsibility for a function, it can be a department's ticket to budget appropriations.
- The source department may not regard other departments as capable of properly using and maintaining the data.
- Personality conflicts, institutional conflicts.

It must be in the interest of the departments to share. Either they will benefit from information products they are now able to make, or they need to demonstrate their compliance with instructions from on high (treasury board directive, cabinet agreement, government policy directive).

With the proliferation of the community GIS movement, data sharing has increasingly become the norm. It is now extremely simple to locate, explore, and download vast quantities and types of GIS data. This certainly makes data acquisition easier today than it was a few decades ago, but don't throw all caution to the wind. It is your responsibility to verify the accuracy of the data and ascertain its legal constraints. Just because you can download a dataset doesn't necessarily mean you may use it, especially for commercial gain. If you are publishing a map using data produced by another organization, it is good practice (and in some cases, legally required) to credit the source.

Legal review

Your final report should recommend a legal review if legal issues regarding the data use and liability become apparent. For example, even though you won't be using the original paper documents again after automation, you may be required to preserve them if they contain legal descriptions for land parcels in the United States.

In the field of surveying, original paper maps may have to be retained as a legal requirement. To date, digital data does not have the same legal status as the written record, and in the event of a legal dispute, it is the paper map that will be used. This may change as surveyors adopt new technologies.

During preparation of your implementation strategy you should check with the acknowledged legal experts closest to your situation. The rapid acceleration in the use of GIS data by government and other institutions is raising many new and interesting legal questions related to the transition of maps from paper to digital forms. Realize these legal issues are out there and make sure any that pertain to your GIS are thoroughly investigated before deciding how they might factor into implementation planning.

You should also recommend measures to reduce the risk and liability associated with possible errors in your organization's GIS data and information products. There are various types of error in data, and it is important to understand your organization's liability and recommend actions to limit that liability. Erroneous data may result from faulty data-input methods, human error, or programs or measuring devices that are used incorrectly. Some errors are more serious than others, and the resulting liability can affect other applications. GIS managers need to know the extent of their liability.

Security issues

You must conduct a security review to determine the type and amount of security necessary to protect your GIS from damage. Through an appraisal of the security review, for your final report you can make recommendations on how to reduce the chance of system damage and data corruption.

Traditionally when referring to *security*, the primary thought in mind has been protection of your investment in data. Sadly, terrorism concerns are with us these days, too, bringing a more layered meaning to the word. In such an atmosphere, off-site data duplication and functioning system capabilities become more important. Some agencies need zero or minimal downtime arrangements in order to provide "business continuance."

A secure GIS provides *confidentiality*, *integrity*, and *availability*, all of which are known as the "CIA security triad." You want to be sure sensitive data, applications, and information products remain secure and accessible to only those who need them; that they remain unharmed; and they are where you need them, when you need them, with backup solutions in case of a security failure.

The concept of security in depth is helpful. Although no security solution is perfect, a layered defense is most likely to approach the degree of protection necessary. Every level of the architecture should be protected by the appropriate access control, beginning with controlled access to the desktop environment. Consult the acknowledged experts first to see what approach to reviewing security issues is advisable in your organization. An example of layered defense is shown in figure 11.5.

Protecting your GIS investment

System backups and data security are designed to protect your GIS investment. One of the major GIS benefits that come to organizations is the increasing value of their databases as they grow over time. Whatever the current value of your database, if it is properly maintained, its value in five years will probably increase dramatically. The value of information derived from the GIS database increases due to the improved business processes the GIS functionality delivers. The successful GIS will often quickly become an integral part of an organization's daily operations. This is why a sound security plan is required.

The risk of your system or data being destroyed or somehow compromised is real and deserves serious attention. Any information system is vulnerable to both deliberate and accidental damages. A disgruntled employee might purposely corrupt data, hackers may steal information, a computer virus could find its way into the server through e-mail, or terrorists could destroy the buildings involved. Natural disasters also pose a threat. Earthquakes, fires, hurricanes, and lightning are all examples of natural hazards that could disrupt a GIS.

While deviant behavior and natural disasters are intriguing subjects, threats more common to many organizations are found in day-to-day operations.[2] Consider the potential effects of coffee spilled in the wrong place, a well-intentioned employee who accidentally deletes or corrupts a database, or a power disruption with no automatic battery backup.

Conducting a security review

When conducting a security review, examine the physical, logical, and archival security of your databases (figure 11.6). Physical security measures protect and control access to the computer equipment containing the databases. Physical security guards against human intrusion and theft (security doors, locking cables) and environmental factors such as fire, flood, or earthquake (fire alarms, waterproofing, power generators). Here are some recommendations for physical security:

2 According to a 2010 CSI Computer Crime and Security Survey (`http://gocsi.com/survey`), the top five reported computer attacks are malware infection, phishing fraud, laptop or mobile hardware theft or loss, bots/zombies within the organization, and insider abuse of Internet access.

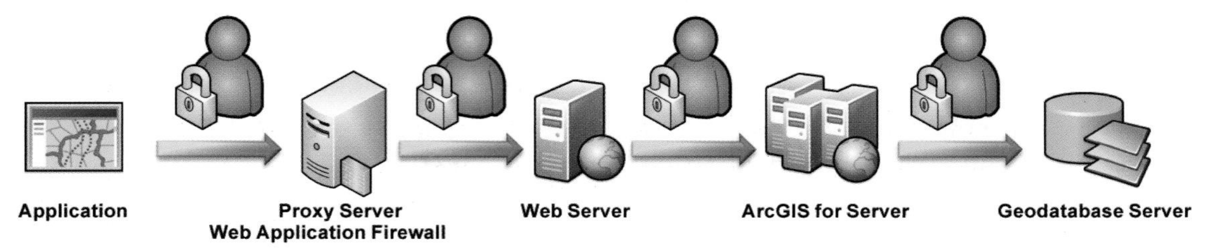

Figure 11.5 **Layers of security across your enterprise.**

- Restricting access to the room in which main data storage terminals are located
- Reviewing construction plans for new stations as available
- Installing fire and intruder protection alarms
- Implementing document sign-out and follow-up procedures

Logical security measures protect and control access to the data itself, either through password protection or network access restrictions. A common security measure is to specify that only database management staff have editing and update rights to particular datasets. Here are some additional ideas on logical security:

- Develop a policy for terminal access
- Protect and control all storage media
- Develop a schedule for virus scanning
- Create a matrix showing access by document type

Archival security means ensuring that systems are backed up and that these backups are stored correctly in remote off-site locations. Legally many organizations are required to archive their data. This means that the system must include functionality that creates these archives. Raw data transfers are not enough: metadata, information about past coding and updating practices, and the location of data must all be stored to allow for quick recovery in the event of a system failure. Consider the following ways to ensure archival security:

- Establish an off-site facility or cloud-storage service (private or public) to store archived data
- Establish an audit trail to track copies of datasets
- Capture every transaction

After your initial security review, consider how current and future security risks might affect what you recommend in the final report. These recommendations

Physical security	Logical security	Archival security
Prevent access to main data storage from the back stairs.	Develop a policy for terminal access.	Establish an audit trail for copies of data.
Review the construction plans for the new building to ensure appropriate climate control.	Create an access matrix by document types.	Establish an off-site backup facility.
Build a public workroom so staff members don't have to go into the vault room to do their work.	Review protection of storage media.	Create and organize metadata.
Initiate document sign-out and follow-up procedures.	Purchase antivirus software.	Purchase storage media.

Figure 11.6 **A typical security review.**

should take into account physical, logical, and archival issues related to security.

Enterprise security strategy

Large organizations should be sure to establish a security protocol that is followed by all departments, at all levels. A proven workflow is shown in figure 11.7.

Once you have determined acceptable risk, you should review known industry threats and assess your current security environment. Be sure to consider threats to datasets, information products, applications, and systems. Look for patterns in your current environment—if there is a problem in one area it may be pervasive throughout the organization.

Being fully informed about the latest security trends will help you to intelligently review options and define your organization's security requirements. When it is time to design, build, and/or purchase the components of your security solution, consider enterprise-wide security mechanisms as well as application-specific options. When the security strategy is implemented, upper management needs to account for any extra staff resources required for security system operation and support. Security costs may be added to the GIS department's budget or the overall enterprise, depending on the organization and its unique solution.

Finally, security defense must be regularly measured and re-assessed for adequate protection. Is it working? If there were some security failures, were they acceptable or does the security solution need to be adjusted?

Existing computing environment

You have already outlined in conceptual form the infrastructure needed to support the new GIS. Review your conceptual system design for technology (chapter 9) before finalizing it as part of the implementation strategy. Your organization may have significant time and money invested in the existing technology base (see "Migration strategy," in chapter 10). Your implementation strategy recommendations should take your organization's hardware and software preferences into account.

The GIS that you propose to implement almost undoubtedly represents new technology and processes in your organization. Very rarely will your planning process lead to a system that is completely new from the ground up. Instead, the typical GIS plan must consider how the GIS will integrate with the existing computing environment, the so-called legacy systems.

As their name implies, legacy systems bring along a history of usage, and people will continue to rely on them. Clean connections to legacy systems are crucial to overall project success. Written organizational standards and policies on system-integration issues related to the GIS will help smooth the implementation process, reduce resistance to a new computing system, and build support among end users.

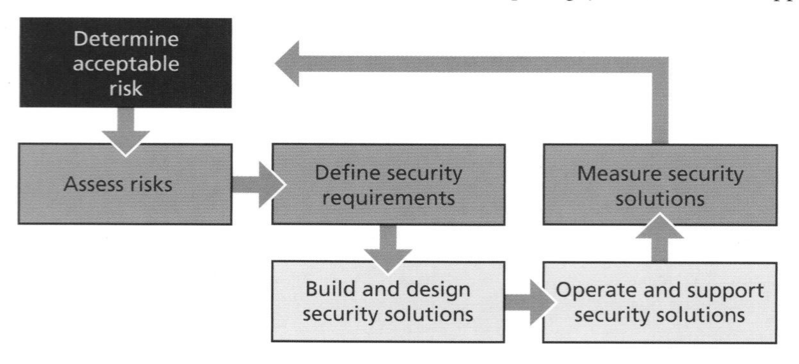

Figure 11.7 **Enterprise security strategy workflow.**

System-integration issues

By now, you've already considered the following aspects of the existing computing environment in which your GIS must coexist; now revisit them and optimize any needed upgrades or changes to hardware and software already in place.

Current equipment: List the current vendor platforms in use across your organization (including remote sites).

Layout of facilities: Gather or create diagrams of facility locations and their associated network facilities (by department and site).

Communications networks: List all the types and suppliers of networks intended for use by the GIS link (dedicated, in-house, commercial). Understand what's happening on both the local and wide area networks. What are the protocols: TCP/IP, IPX, NFS? What are the bandwidths for each link that will be used by the GIS (e.g., T3, T1, ISDN)?

Potential performance bottlenecks: Identify missing or inadequate communication links, fault tolerance, system security, and response-demand issues.

Organizational policies and preferences: Consider the IT culture of your organization. Is all hardware from a single preferred supplier? Is there a standard operating system in use? If policies for procurement and standards for use have been adopted or maintenance relationships established, consider these carefully. Adopting different systems may lead to problems with support and acceptance and to a need for additional staff training.

Future growth plans and budget: The GIS must work within the fiscal framework of the organization. Budgets set must be adhered to, so the GIS manager must always keep an eye on costs and ensure the work is efficient and affordable. If money runs short, things like breadth of user applications, system performance, or reliability will take a hit. Initial budget projections should be considered in later benefit-cost analyses.

Migration strategy

You have examined the existing computing environment of your organization. Now you must recommend a migration strategy for moving from the existing system to the new GIS. Include with this strategy detailed plans and a timeline for merging the new system with legacy systems. Recommendations should address whether or not to replace, rebuild, or merge the new system with modeling techniques used by any legacy systems. A legacy transition workflow might look something like this:

1. Build a pipeline to new environment with scheduled refresh for all framework data.
2. Systematically build new data framework with associated maintenance procedures in place.
3. Link legacy applications to new data framework environment.
4. Either port or build new business applications based on a formal assessment of product life expectation.
5. Systematically phase out legacy applications.

Risk analysis

Implementation can be affected by four major groups of risks: technology, budget constraints, project management and scheduling, and human resources. Recommend steps to mitigate the risks identified. Include the results of the risk analysis in the final report to help ensure that senior management is aware of any implementation difficulties identified beforehand. (Review the risk analysis section of chapter 10.)

Alternative implementation strategies

Include alternative strategies for implementation in the final report. Recommending more than one strategy (even if you have a strong personal favorite) shows that you're looking for the best approach for your organization. Consider using pilot projects with a subset of the geographic extent, particularly if you're concerned about the ease of integration of any particular datasets. Beware, however, of using a pilot project as a replacement for planning. That approach inevitably leads to an incomplete system design and a certain measure of frustration and wasted time.

System procurement

How you will procure your system is part of implementation planning. You must consider two key factors: the procedural requirements of your organization and the characteristics of the planned system.

Many organizations have purchasing requirements that become more rigorous as the expected expenditure increases. Low-cost, generic products can often be purchased with minimal fuss. Acquisition of more expensive products and services may require most, if not all, of the following steps leading up to decisions about procurement.

This list represents the most elaborate procurement process you could encounter; the average organization might include only half of these steps. Whatever your organization requires, by listing all of these steps in the final report you show that you have accounted for those required during your acquisition planning. Follow the steps that apply when it comes time to procure your hardware and software technology and services.

Step 1: Request for qualifications (RFQ)

The RFQ is a request for qualifications from each potential technology vendor. The RFQ is optional but done early in the procurement process. In response, the vendor should identify the types of systems it has provided along with a track record of experience. This will help you determine which vendors qualify to provide systems similar to the one that you plan.

Step 2: Request for information (RFI)

The RFI is particularly important if you have a large procurement that could be so coveted by vendors as to result in potential protests from vendors (see Focus box on the following page).

Step 3: Request for proposals (RFP)

The RFP invites vendors to propose the most cost-effective combination of hardware, software, and services to meet the requirements of your organization. (Detailed guidelines for how to prepare an RFP appear in the Focus box on page 189–190.) In your RFP, ask the vendors to put their proposals into a format that you specify. This will allow you to develop a rating system for proposal comparisons. If you are procuring a major enterprise-wide system and benchmark testing is planned, let the vendors know that their proposals will face these tests.

Step 4: Receipt and evaluation of proposals

Identify who in your organization will serve as evaluators and spell out the criteria that they'll be using to make that evaluation. These criteria are often a key part of the RFP.

Step 5: Benchmark test

Benchmark testing involves verification of the proposals from the top-rated vendors to ensure they can perform the tasks and functions necessary for the planned system and communication network (see appendix B). Vendors may not wish to participate in a benchmark effort if you request extensive testing but are not planning to make a relatively large acquisition. In the event that thorough benchmark tests are not carried out, the purchaser should request a demonstration of system capabilities that includes the full functionality required for the organization. Always, caveat emptor—let the buyer beware.

Step 6: Negotiation and contract

Negotiation and contract often involves cooperation with your purchasing and legal staff. Personnel from these two departments can really help during contract negotiations, but as the manager, just make sure you review the contract to ensure that the technical requirements are properly addressed.

FOCUS

Request for proposal (RFP) outline

Early in your implementation planning process, you may want to write a request for proposals (RFP) and send copies to hardware and software vendors in contention for your business. An RFP invites vendors to propose the most cost-effective GIS solution to a specified business need. Well-crafted RFPs are not shopping lists of hardware and software to be procured, but rather descriptions of the work that has to be done by the system. An RFP also specifies how the selection and actual procurement will be carried out, so that vendors can propose realistic options based on the most complete information.

Sometimes the actual RFP may be preceded by a request for information (RFI), which many organizations use to solicit assistance in the development of their final RFP. Savvy technology managers realize that early input from tech suppliers can be very instructional—vendors spend a lot of time thinking about how to implement their particular technology. Sometimes an RFI is nothing more than an early draft RFP. RFIs are particularly useful when the scope of your system is likely to be large. You can include the draft RFP when you send the RFI to vendors, asking them to respond with a letter of intent to bid and any comments. This step helps you to design an RFP that is not inadvertently biased toward one vendor or another.

The RFP for a major procurement is typically a single document supported by substantial appendixes. (For a smaller project, the single document may be all that is needed.) The primary document lays out the requirements, while the appendixes provide details, such as the master input data list, the information product descriptions, copies of government contract regulations, worksheets for product cost and data conversion estimates (to standardize replies from vendors), and overall data processing plans pertinent to the GIS.

This primary document of the RFP should include the following:

General information/procedural instructions: Here, you cover the acquisition process and schedule, the vendor intent to respond, instructions on the handling of proprietary information, proposed visits to vendor sites and debriefing conferences (if deemed necessary), clear rules for the receipt of proposals (date, time, place), and arrangements for communications between the organization's contact person and the vendors.

Work requirements: This section is the most substantial part of the RFP. It should list the work to be done by the system, expressed specifically in the context of the information products that have to be created, the system functions required to produce the information products, and the data needed in the database. Limit the main text of the RFP to a listing of what is required, and leave supporting definitions and any necessary descriptions to the appendixes.

(continued on next page)

FOCUS continued

Here's a checklist of work requirements that every good RFP should include:

- The full set of information product descriptions (IPDs)
- The master input data list (MIDL)
- Data-handling load estimates over time by location
- Functional utilization estimates over time by location
- Notes on existing computer facilities and network capabilities
- Lists of any special symbols required (for maps, etc.)

Services to be performed by the vendor: Typically, vendors are contracted to supply and install the hardware and software systems and to train the new users. But sometimes in-house staff takes on part of that work. In any case, clear, complete written documentation in the form of user guides must be created. After installation and start-up training, also plan for ongoing maintenance and upgrades to the system. Spell out what you want the vendor to do.

References: Good vendors are more than happy to provide reference sites of existing customers running similar systems. After you receive the references, call the other customers yourself to hear their experiences.

Financial requirements: You will be making some significant financial decisions, including whether the hardware equipment will be leased, lease-purchased, installment-purchased, or purchased outright, and depreciation and tax considerations may come into play. The RFP should solicit the cost of the recommended approach along with several alternatives and their respective costs. It should include the instructions on use of financial worksheets, if any, and the licensing terms and conditions on purchased data and software. If you require estimates of operating costs from the vendors, be sure they account for staff hourly rates. Maintenance on software should be a guarantee of prompt upgrades to new versions and technical support from the vendor, so you want to ask for that in writing, too.

Proposal submission guidelines: Describe the required proposal format, the number of copies needed, and the proposal content, including appendixes. (Choose a format that will facilitate your comparing and rating proposals from the various vendors.) Specify any mandatory contractual terms and conditions, or any specific financial requirements.

Proposal evaluation plan: This part of the RFP describes the structure of whatever committees you will use to evaluate the proposal. With the RFP, step through the evaluation process and lay out the criteria that will be applied in the final decision making. These are the rules of engagement, as it were. Take care to explain the procedure with clarity—you want all minds in agreement, within the organization and among the vendors, on the process to be followed. If the system will be subjected to benchmark and acceptance testing, spell out the testing processes. This is particularly important in major procurements; most organizations contemplating major procurements have firm procedures for these steps. Both the organization and the vendors have a stake in your producing a clear, explicit RFP.

Step 7: Physical site preparation

Make sure that the site is properly prepared before the system is installed. This seems like an obvious point, but it can cause problems if overlooked. Are the necessary servers and appropriate network connections available for system installation? Does each GIS team member have a chair and a desk?

Step 8: Hardware and software installation

Specifying who will install the system's hardware and software depends on the complexity of the system. In some cases, the vendor will provide this service. If your organization has the technical staff available, you could install the system in-house. Basic PCs have evolved a long way. It's possible to take delivery of a boxed computer and have it set up and running in less than an hour. Often, the vendor and the client are both involved in system installation.

Step 9: Acceptance testing

Within the RFP you would have specified the methods chosen for acceptance testing after system installation. These tests often require the hardware components to operate without error for a given period of time. With many of the widely used components available today (e.g., personal computers and their operating systems) these tests are of limited added value. When planning acceptance tests, consider testing the system's ability to integrate with existing databases and software and test all these connections. Network connectivity problems are always a major source of bugs and should be tested thoroughly for possible problems.

Activity planning

Activity planning is a specific plan for information-product delivery.

If you've followed the procedures outlined in this book, by now you should have a clear sense of the following:

- Information product priorities (perhaps modified somewhat by multiple-data-use potential)
- Data priorities
- Data-input timing
- Application programming timing

FOCUS

Selection criteria

The final choice of your GIS is based primarily on its ability to perform the functions specified to create the information products. This is the first litmus test of whether the system is acceptable. Other factors that factor into the selection criteria include cost, training availability, system capacity and scalability, system speed, system support, and, last but not least, vendor reliability (i.e., their financial stability, position in the marketplace, and verified references).

Having already prioritized and classified the functions based on their frequency of use and relative importance to total system functionality, you can use this information as selection criteria. Also use the rating system you've developed for proposal comparisons.

In your recommendation for technology acquisition, be sure that the equipment to be procured will be sufficiently used from the outset. Optimum use should be made of the equipment capacity acquired in the first year. It follows that it is cost-effective to have a continuous technology acquisition budget to maintain your system. You should recommend the allocation of resources to keep your system cost-effective (see "Acquire technology" on page 200).

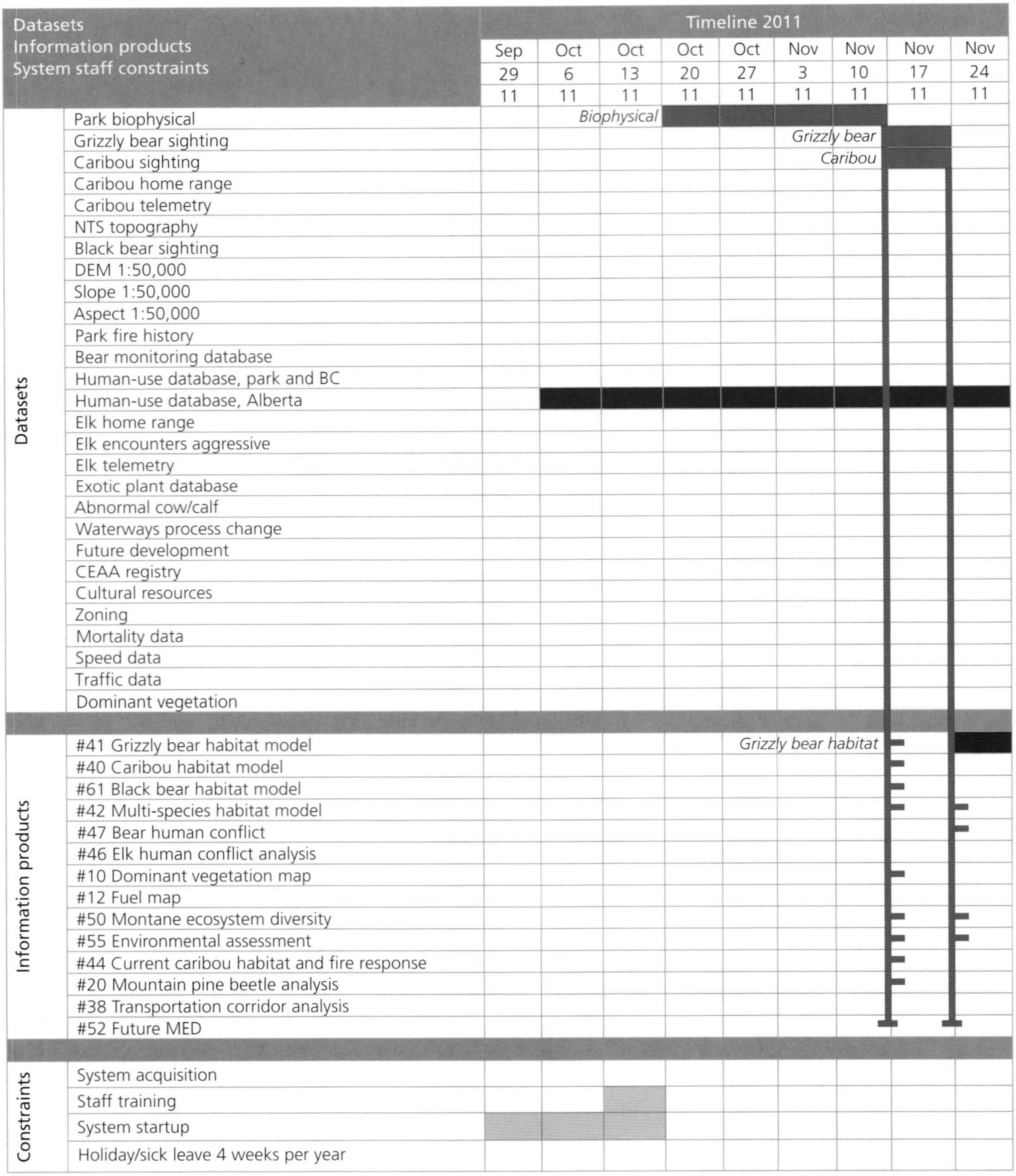

Datasets Information products System staff constraints		Timeline 2011								
		Sep 29 11	Oct 6 11	Oct 13 11	Oct 20 11	Oct 27 11	Nov 3 11	Nov 10 11	Nov 17 11	Nov 24 11
Datasets	Park biophysical			*Biophysical*						
	Grizzly bear sighting							*Grizzly bear*		
	Caribou sighting							*Caribou*		
	Caribou home range									
	Caribou telemetry									
	NTS topography									
	Black bear sighting									
	DEM 1:50,000									
	Slope 1:50,000									
	Aspect 1:50,000									
	Park fire history									
	Bear monitoring database									
	Human-use database, park and BC									
	Human-use database, Alberta									
	Elk home range									
	Elk encounters aggressive									
	Elk telemetry									
	Exotic plant database									
	Abnormal cow/calf									
	Waterways process change									
	Future development									
	CEAA registry									
	Cultural resources									
	Zoning									
	Mortality data									
	Speed data									
	Traffic data									
	Dominant vegetation									
Information products	#41 Grizzly bear habitat model							*Grizzly bear habitat*		
	#40 Caribou habitat model									
	#61 Black bear habitat model									
	#42 Multi-species habitat model									
	#47 Bear human conflict									
	#46 Elk human conflict analysis									
	#10 Dominant vegetation map									
	#12 Fuel map									
	#50 Montane ecosystem diversity									
	#55 Environmental assessment									
	#44 Current caribou habitat and fire response									
	#20 Mountain pine beetle analysis									
	#38 Transportation corridor analysis									
	#52 Future MED									
Constraints	System acquisition									
	Staff training									
	System startup									
	Holiday/sick leave 4 weeks per year									

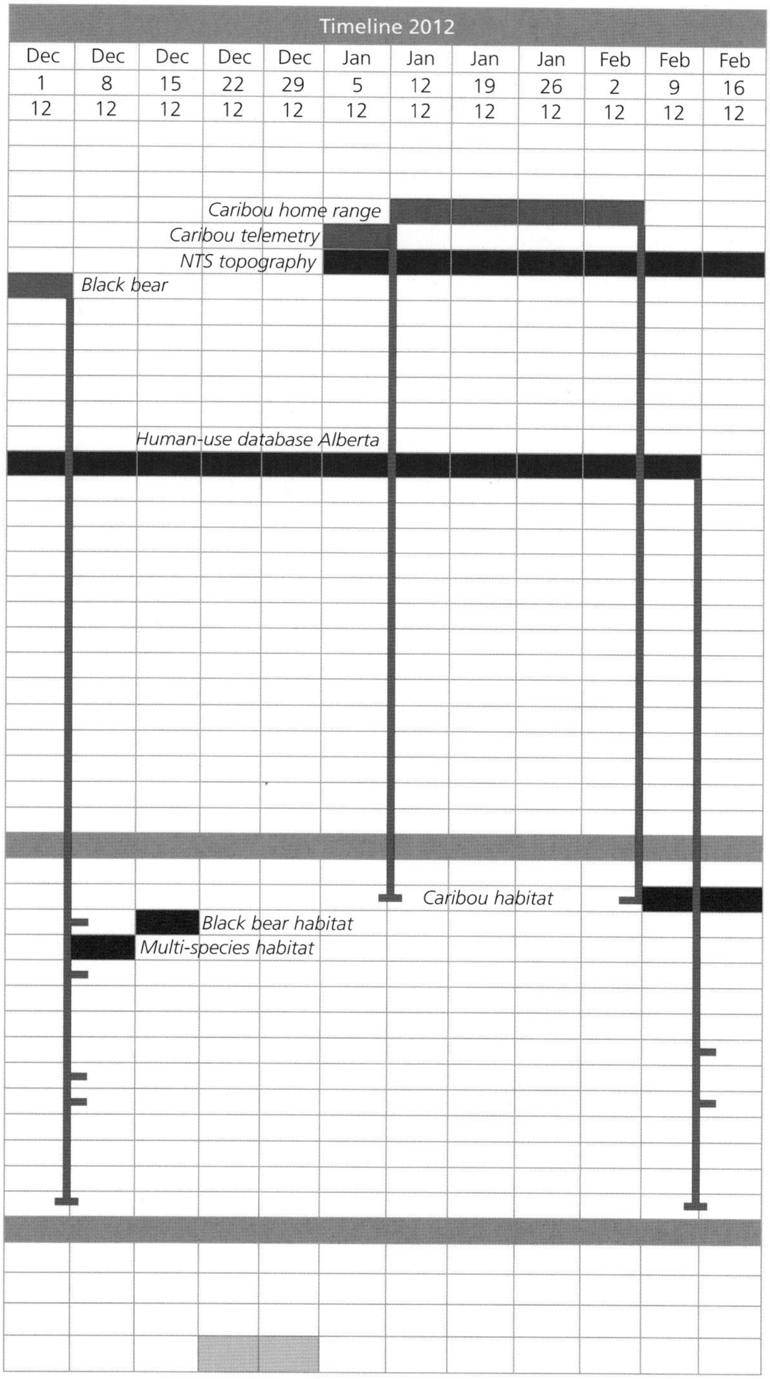

Timeline 2012

Dec	Dec	Dec	Dec	Dec	Jan	Jan	Jan	Jan	Feb	Feb	Feb
1	8	15	22	29	5	12	19	26	2	9	16
12	12	12	12	12	12	12	12	12	12	12	12

Caribou home range
Caribou telemetry
NTS topography
Black bear
Human-use database Alberta
Caribou habitat
Black bear habitat
Multi-species habitat

Figure 11.8 **Gantt chart showing a national park's GIS planning for input of datasets (red and green horizontal bars), readiness of datasets for use (blue vertical lines), availability of information products (purple horizontal bars), and system staff constraints (yellow).**

- System acquisition timing
- Training and staffing timing

Now you are ready to start the overall activity planning for your GIS. First, revisit the information product priorities in order to refine them by answering these questions:

- Will some information products depend on the creation of other information products? If so, the latter must be assigned a higher priority.
- When will it be possible for you to make each information product, given your estimated data-readiness dates (which include both data availability and the time required to input the data)? This is particularly important for the high-priority information products you want to make early.

Use these considerations to revise your information product priorities and establish a second-level priority review.

The next step in activity planning is to use a Gantt chart—essentially an illustrated timeline with dependencies plotted on a linear chart so that one can see which activities are time-sensitive. Gantt charts, like the one in figure 11.8, can help you plan and manage your project effectively. There are many other project evaluation and review techniques and specialist project-management software packages, but this is one of the simplest and easiest to apply. You need to manage your project so that you focus on the tasks that produce results.

Displaying information about activities and their duration, a Gantt chart allows you to schedule and track the activities associated with GIS implementation. Each dataset or information product, system acquisition, or staff activity is a row in the chart, and time periods are columns in the chart. Bars indicate the time necessary to complete each activity, for example, data-loading application, writing break-in period, or staff holiday. Gantt charts make it easy to see when a particular activity will be completed and when other activities relying on its completion can begin. For example, the entry of a particular dataset might be one activity, at the end of which it would be possible to produce a particular information product.

You can use a Gantt chart to document project activities and their duration, establish relationships between your activities, see how changes in the duration of activities affect other activities, and track the progress of your project. You will also use the Gantt chart to schedule hardware, software, and network acquisitions over time, after these have been determined.

Gantt charts can be constructed using ordinary spreadsheet software (such as Microsoft Office Excel) or specialist project-management software from a variety of suppliers. In relatively simple situations, a handmade one will do.

In the left column of your Gantt chart, first list the name of every dataset you intend to put into your system, followed by the names of the information products you intend to generate in the same time frame. At the bottom, list the constraints—system or staff downtime where nothing can be done in terms of creating information products. These bottom four rows allow time for system acquisition, staff training, system start-up, and holidays and sick leave.

The columns on the top of the chart are usually best divided into one-week segments grouped into months and years. You should indicate fiscal years or budget cycles. You can use vertical dotted lines for this. Note that, while our sample is just a "snapshot" of the time frame, a Gantt chart for a GIS project typically covers several years.

When you start charting, you are most likely to place activities onto the Gantt chart in the following order:

1. System acquisition and staff training activities
2. Data-input activities—establish when data will be ready
3. Application development activities for any high-priority information product

The order reflects the logic of your planning. Having established when the system will be installed and staff available, you can coordinate the timing of the input of the data with the priority of need for information products. Acquired (bought) datasets tend to require more time for input than datasets generated in-house. Usually an in-house dataset (appearing in red on the chart) can be made ready more quickly because it originates with the organization and is less likely to require conversion or extensive reformatting. In this case, at the top of figure 11.8, a national park is collecting data about grizzly bear and caribou sightings, two datasets being prepared for use in the space of a week. The park will also buy, or otherwise acquire, datasets like the human-use database Alberta charted in green farther down the list. Work may be needed to make datasets from outside the organization fit the GIS, thus the green lines on this chart may span many more weeks than the red.

Meanwhile, in the category below the list of datasets on the far left, information products await this data readiness. The vertical blue lines on the chart indicate when the datasets they are attached to will be ready for use in creating information products. The "tics" along the blue lines mark the information products (from the column on the far left) that will use a particular dataset. The rows at the bottom of the chart show time dedicated to training, statutory holidays, and so on, when staff or system will not be available.

Sometimes what's called an *interim information product* is generated on the way to making another information product, which can be useful by itself or frequently as a dataset for yet another information product. You can plot this out, too: the purple bars next to the blue tics on the chart signify this potential of multiple use. Also occasionally, after you've got the first ten or so datasets into the system, you realize you could make some information products of less priority sooner than those of high priority. This is called *multiple data use,* and it affords you the opportunity of adjusting your priorities in order to make the best use of what you can create right away.

Gradually, you build your Gantt chart, taking into account restrictions of staffing and product demand, until you have decided how and when to build each information product. Now you have, for the first time, a date on which you can reasonably anticipate the delivery of the first of each information product. This plan for information product delivery should be part of your final report that you present to secure approval for implementation.

The final report

While addressing all the aforementioned issues that apply to your situation, you are developing the GIS implementation strategy you will present to executive management. These are the choices you have made, the actions you are recommending, the timetable for your implementation plan. These recommendations will be a key component of the final report and presentation you will give to the board, after approval from the management committee. After the management committee reviews the final report, you will note any extra implementation actions now required or any new concerns identified during the review process, then add them to your final recommendations.

Report components

In constructing the final report, lay out your recommendations for GIS implementation and what you've done and discovered that has led you to these conclusions. This is why you've done the work—to win approval to get a needed system in place—so make it good. The report itself will probably be anywhere from fifty to a hundred pages in length, with the executive summary section at the front (though you will write it last). It will be supported by a set of appendixes full of all the technical information you've generated.

The subsequent presentation of your recommendations in front of the executive board should also be of the highest quality. Go over the entire project and tell them what you've been doing. The beginning was all about finding out the strategic goals of the organization, its way of doing business, and the information products necessary to help move the organization toward these goals. You found out what data would be needed to produce that information, the software required to get it, and the hardware and communications infrastructure to support it. You projected the costs in your benefit-cost analysis, and then asked yourself how do we get that in place? What's going to stop us from implementing this? One by one, you came up with ways to overcome obstacles like the knowledge gap. Realizing that a trained staff is paramount to success, you thought about staffing and training, and now you're recommending exactly what the organization needs. Don't drag out the work that's been done to back up these recommendations. It stays in the ring binders in the cardboard box. It's enough to say what it contains.

Do tell the board enough in your presentation and in the pages of the final report so that they can approve the implementation plan. They need the specifics in front of them, such as the timeline—"you need two extra staff positions in budget A by a particular date"—and so forth.

The final report is composed of these six components, which follow the executive summary:

1. Strategic business plan considerations: A synopsis of the mandate and responsibilities of the organization—a summary of what they do and their business model for success. This leads to understanding what they need to know and how GIS and the information products will be effective helpmates in the strategic effort toward realizing these goals. Remember that GIS means change. In an enterprise, because its implementation may alter the way they do business, it becomes particularly important to keep the focus on how GIS serves their business. In executive terms, this is a mind shift. They need to know why you're recommending they go to a cloud-/server-based system or not, for example. You will have kept them informed every step of the way, of course, so the executives will already have a feeling for what you're proposing because they will have contributed to establishing these terms. You simply need to be sure to clearly link your recommendations to the organization's goals, so that in approving your plan the executives are confirming their shared mission.
2. Information product overview: A brief introduction to the information products that will be created and the datasets that must be acquired to generate them. List the information products by their titles and groupings, along with the datasets they require, also listed by name. You won't include the MIDL here, but you will say that all its data, as well as the full IPDs, are in the appendixes of your report. Make it clear how each information product supports the overall strategic business plan.
3. Conceptual system design: the synopsis of the work on technology. You have identified the most suitable data design and software/hardware system design and the communications infrastructure necessary to support it. Show how you came to these conclusions. Why did you choose this particular data model? How does this specific client-server architecture best meet the needs of the organization?
4. Recommendations: Your recommendations point out the most direct path forward for putting GIS in place. This is the implementation plan you are proposing that has been approved by the management committee. Your understanding of the organization's needs and what's required from GIS to meet them has led you to formulate these specific, practical recommendations. Be sure to include the benefit-cost analysis here. Don't neglect to recommend a migration strategy for moving from the existing system to the new GIS. When you are recommending alternative implementation strategies, you may want to include a pilot project as one of the options, but be careful how it is used.
5. Timing: Set forth the dates that show the timeline for your recommendations. Use a Gantt chart to show the milestones.

6. Funding alternatives: Detail the budget for all this. Here you are saying the funds for one thing will come from budget A and the funds for something else from budget B. There may be external sources that will allay these costs such as grants or data-sharing arrangements. Recognizing that managers worry about the financial effects of technological change, you should emphasize that technology is getting cheaper and faster. Maintaining out-of-date equipment is a waste of funds, and so is buying beyond capacity. It's better to spend a little every year to keep up with technology; about 20 percent of your total acquisition budget is a guideline that works. Technological change is going to happen, so you manage it with incremental budgeting and an understanding of system life cycles.

After completing these sections, write the executive summary section and place it at the front of the final report. After one more review by the management committee, give the report to executive management and prepare to put on a dynamic presentation to the board. This presentation may be a formality, but it is an important one. Remember this is affirmation time, a chance for the board members to give the formal approval that makes the implementation happen.

GIS management committee review and approval

The information prepared during the planning process is the basis for your recommendations, and the management committee should already be fully conversant in it. After thoroughly reviewing your implementation strategy (and the planning materials backing it up), the committee will approve, revise, or reject the recommendations. If you've done your homework, the approval should come quickly and with renewed enthusiasm for the mission. Never rely on a consultant to do this review. It must be undertaken by staff inside the organization, and preferably by a group and not an individual, to reap the strength of combined views.

The GIS management committee can, of course, adjust the implementation plan. They should also provide ongoing support during and after implementation. This committee is an integral and permanent part of the GIS team effort. The members might change, but the committee will always be needed.

Merging the GIS plan with the overall business plan

Once the implementation recommendations have been approved by upper management, they should ideally be folded into the organization's overall business plan, particularly the management accountability framework. This ensures the GIS project is not just a peripheral experiment, but a core vehicle for achieving the organization's mission. The GIS manager should make this recommendation once the implementation plan is approved, but it is up to the management committee to make it happen.

Implementation change

The business model of the modern organization is characterized as dynamic. Implementing a GIS to fit this model is a continuing process of adjusting to change, since it occurs in an ever-changing environment. There are changes in technology, both in hardware and software. But there are also sometimes subtle, sometimes profound, changes in the business needs of your organization and perhaps in the institution itself. As institutions shift gears to keep up with advancements in technology, the knowledge gap widens, which in turn signals even more challenge in the areas of staffing and training. So much change needs to be managed. Managing change starts with understanding the types of change.

Technology change

If you don't think technology is changing GIS, consider that over just the past fifteen to twenty years, the hardware used for GIS has evolved from mainframes to minicomputers, from minicomputers to workstations, from workstations to personal computers, and now to handheld devices. The use of servers—in-house or cloud-based—as computer centers linked on a network is increasing significantly. Operating systems have changed from being proprietary and hardware dependent to hardware independent. To accommodate these changes, GIS software has constantly evolved.

Change is generally a good thing when it comes to technology, as we began to explore in chapter 9 with the latest figures on the rate of technology change and the life cycle of technology. New versions of software and hardware do actually make the work easier and more cost-effective. Newer hardware brings significantly faster performance and cheaper storage, making work easier. New versions of GIS software offer easier use, simplified procedures, and reduced repetitive work. New user-friendly versions tend to fix bugs, becoming less error-prone and more stable and reliable.

The swift rate of change in technical capability also brings its share of challenges. Advancements in technology move faster than the changes in institutional requirements, and to meet this challenge you must employ the criteria of cost-effectiveness. Rapid advancements in technology mean that maintaining older software and hardware can become prohibitively expensive in less than five years. Vendors must offer maintenance for new hardware and software. Over time, they increase the price for maintaining older technologies to focus more of their resources on the latest technology, encouraging the migration to newer versions.

Institutional change

The business needs of most organizations change incrementally over time. The staff gains more experience, new ideas and approaches are brought in, and new information products to support the new business needs are requested. This is the natural growth cycle of a successful business operation. Incorporating the new information products into the GIS workload will present no significant problem, if you have planned for growth in advance.

Occasionally, major institutional changes occur, such as the merging of companies or departments of government, major changes in mission for all or significant parts of the organization, or changes due to recession or other budget constraints. In such cases, it may be that one organization has done no GIS planning, while the other is quite advanced. Regardless, a new overall suite of information products will probably be required, involving new datasets and database designs. In this event, you should review the entire GIS plan and put in place new enterprise-wide planning, considering all the new ramifications for benefit-cost, technology acquisition, and communications that come with it.

The rates of change in technical capability and change in institutional business needs are illustrated in figure 11.9. The differences in the rate of change are dramatic, but they can be managed. The challenge for the GIS manager is to provide the maximum business support while minimizing expenditures. He or she manages the changes in technical capability on the basis of cost-effectiveness. When new information products are introduced in response to changing business needs, this is managed by maximizing support for the business of the organization.

In the early days of GIS, plans were typically updated on a five-year cycle. These days five years is too long; things change too fast. To determine when your implementation plan will become outdated, you must carefully assess the technology changes that have occurred since preparation of the last plan. When updating the implementation plan, the most important factors for you to consider are the following:

- The need for the continued availability of existing information products in the most cost-effective manner
- The expansion of the information "product line" to meet the changing business needs of the organization

Managing change

The best advice for managing change is exactly the same that informed your planning for a GIS from the first step forward: start with the goals and objectives of the entire organization in clear focus. Keep in mind that you will be talking about change whenever you approach management for approval to add information products or acquire new technology, so it is best to update management regularly, just as you keep your plan itself up to date (figure 11.9).

Start with an enterprise-wide plan

Enterprise-wide planning allows for maximizing the benefit that can be gained from GIS. Note that enterprise-wide planning does not mean enterprise-wide implementation. (In the end, it may be prudent to implement only a small part of the plan at first and build on the success of that part for further implementation.) The point is that, while it's usually satisfactory to implement incrementally, it's always best practice to plan comprehensively. When you do, you can identify the most important system functionality on an enterprise-wide basis. You can intelligently consider GIS communication requirements throughout the organization. Time spent in enterprise-wide GIS planning is the best investment related to GIS that you can make because it sets you up for success.

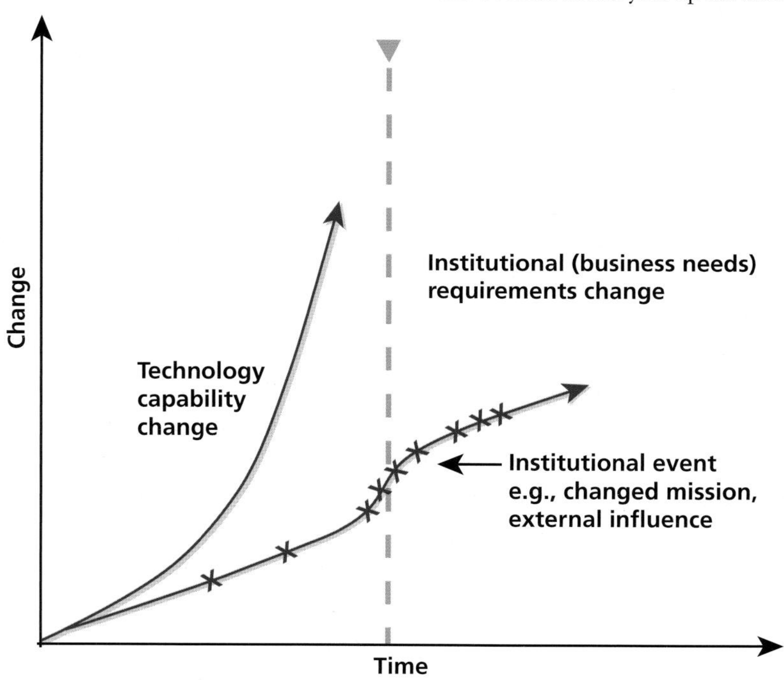

Figure 11.9 **Adding information products (x) to manage major change.**

The wise advice is to plan early and as broadly as possible. But you also need to plan within the normal life of an organization. The one exception to that relates to special circumstances that can be imagined but are out of the ordinary. For example, planning can and should be adopted for one or a small set of very high-priority information products created in response to crises or natural disasters.

Be a bit wary of out-of-the-box software with available data to produce so-called standard or all-purpose useful information products (followed by another and another). With such generic information products, after a while, demands and problems start to arise, and you are forced to actually think about what you are doing. Again, your planning, like each information product, needs to fit the specific mission of your organization.

Acquire technology

These days, you can—and should—use much more of the capacity of the procured technology right from the outset. You should plan to use at least 50 percent of the capacity of the equipment you choose in the first year of its acquisition; otherwise, buy less. The old paradigm of buying the largest equipment possible from the current budget and planning to use 10 percent of its capacity in the first year, 20 percent the second year, 50 percent the third year, and so on over a five-year cycle is no longer cost-effective. Given the increases in equipment capacity and the cost reduction over each twelve-month period, you can do better.

In other words, to transform this type of change into the opportunity it really is, you need to set up a continuous technology acquisition budget to maintain your system cost-effectively, and senior management needs to allocate the resources for this. Hardware, software, training, and data all need to be procured bearing in mind the life cycle of both the technology and the information products.

To do this, describe your GIS as part of the organization's infrastructure. As with other elements of this infrastructure, two types of funding are necessary:

- Funding for ongoing operation and maintenance
- Funding for capital investment

A common approach to ongoing capital investment is to estimate the life cycle of the asset and annually budget for an appropriate amount in a capital replacement (or sinking) fund. For example, three years may be the estimated life cycle for a desktop workstation. If your organization has sixty of these workstations, you should budget funds to replace twenty per year.

In common practice, hardware extends its useful life by cascading down through the organization; older workstations are assigned less-demanding tasks, while high-end workstations are replaced on the acquisition budget. Many organizations also regard initial data acquisition as a capital investment item rather than an operation and maintenance expenditure.

Add information products

Adding new information products after the GIS is up and running is not a problem, but it does require careful management because now you must weigh priority of need for the new against the old. First, you will revisit the steps you took in activity planning to determine the priority level of the new information products just as you did for the original products when you were creating the MIDL. But this time, recognizing the demand new products will place on the data-entry schedule, you will consider whether displacement or delay might affect generating the other information products and their benefits.

In essence, you must formally manage the process of scheduling new information product design and generation in the context of the existing information product priorities, data availability and readiness, activity timing, and revised onset of benefit from other information products.

Information product design, particularly for more complex products, can be a time-consuming process and must be judged in the context of the life cycles of the technology being used. For example, if the technology currently in use is late in its life cycle and has only one year until obsolescence, a major information product design should probably wait and take advantage of the new technology.

If waiting allows for significantly reducing the amount of time needed to develop the product, it could measurably reduce its costs. Alternatively, you could always go forward with the original design, but only with the full understanding that its benefit cycle will be limited; it may face abandonment, or adapting it to the new technologies when they finally arrive may require significant effort.

Inform management about change

Continual change raises a number of questions for the GIS manager:

- How do you secure support for a technology that is constantly evolving?
- How do you convince management that system upgrades are necessary?
- How do you secure funding for training costs?

When facing such questions, keep in mind that GIS implementation will move forward because an organization expects to benefit from using a geographic information system. Never forget your role in educating yourself and senior management about change. Keep up with change—anticipate it, prepare for it, and even make it happen—and your route can lead you straight to the benefits and opportunities change carries with it. Both technology and business needs change with time, and you need to stay on top of it all.

As you will have seen for yourself by now, GIS itself means change—the kind of change that brings benefits—so you are keeping senior management apprised of change when you keep them informed about the benefits received from the GIS as they become known. Your proposals for support required for upgrades and training must be presented in the context of these benefits. Provide management the opportunity to see how the GIS minimizes cost while maximizing support for business solutions. Revisit the benefit categories in your benefit-cost model to remind yourself what specifically to look out for. Identify policy changes that were substantially influenced by information received from the GIS. As demonstrable benefits occur, provide these in a continuing series of reports to senior management. To maintain the cost-effectiveness of your system, you'll do regular budget planning for which you can use these reports as reference.

Keep your plan current

Based on what you know now, it makes sense that the GIS implementation plan will require continual review. It must be periodically updated in response to changes in your organization and to keep pace with technology trends. Consider the planning document to be a "living" document that adjusts as variables—costs, personnel levels, organizational relationships and agreements, and so on—change. Determining when to update the GIS implementation plan requires a careful assessment of the changes that have occurred since preparation of your final report or the last plan. The approach to this is as individual as each organization. It has been found, however, that much of the updating task is amendable to automation, which not only reduces the workload but also provides a structure for updating the document at satisfactory intervals. When changes are made to the GIS plan, it may be necessary to incorporate those changes into the overall business plan, which should now encompass the GIS plan.

Again, as at so many moments during the planning process, when updating the implementation plan, the most important considerations revolve around your information products and how you really need them to be. Ask yourself and your team what needs to be done to ensure both the continuing availability of existing information products for your organization and the needed expansion of the information product line as scheduled. Then make your plan for doing what needs to be done.

As at every stage of the planning methodology, if you ask the right questions, the answers will lead you to the next step in planning a solution. In that way, no matter how much change happens or will occur, in planning you have the means to adjust to it. Fortunately, planning is like thinking: the ability to do it is always there when you need it.

Communication

It is vital that you communicate the results of your labors throughout your organization. It is difficult to overestimate the benefit that good communications have on the subsequent success of the GIS implementation and conversely, the impediments that are encountered in its absence. Sharing success stories (and lessons learned) can benefit your organization and edify the GIS community at large. You might publish an enterprise-wide series of papers or a newsletter that highlights each information product with a clear explanation of the benefits that have and will accrue.

At a broader level, the CEO might present a paper at a national or international conference detailing the overall GIS planning approach, the benefit-cost management paradigm and the cumulative benefits expected from GIS usage over time. Such a paper would be suitable for publication and would reflect well on the organization. You already know how easy and beneficial it is to share your data, maps, and other information products; consider also sharing your GIS story with the burgeoning, connected community of GIS users, to promote methodical planning, to encourage *thinking about GIS.*

Appendix A

GIS staff, job descriptions, and training

A good GIS staff is an invaluable tool for a manager. Money can buy more hardware and software, but even money cannot create the motivation and enthusiasm essential to a successful staff and a successful GIS implementation. It takes people to build, manage, and maintain a GIS, so part of planning for a GIS is making sure you will have enough staff with the appropriate skills and training.

GIS staff

Consider everyone directly concerned with the design, operation, and administration of the GIS as the GIS staff, including the end users, the management group, and the systems administration team. The core GIS staff, however, is where you expect to find the more specialized GIS skills, and it includes the GIS manager and GIS analysts.

GIS manager

The GIS manager requires skills in GIS planning, system design, and system administration. Ideally, the manager would also possess hands-on, technical GIS skills—most effective GIS managers come from the ranks of GIS doers. The responsibilities of this position vary depending on the type of organization. In smaller organizations, the GIS manager may be the only person involved with the GIS; so the same person who negotiates data-sharing agreements with the next county over is also pushing the buttons on the GIS when the data itself shows up. In very large organizations, often the GIS manager is in charge of coordinating the GIS staff, managing site facilitators, working with multiple departments, and overseeing the development of an enterprise-wide database. An emerging trend is toward the creation of geographic information officers (GIOs), who become the true champions and executors of GIS innovation within an organization.

GIS analysts

GIS analysts are people with GIS expertise working in support of the GIS manager. In smaller GIS organizations or in a single-department GIS, the GIS analyst may be one person with a broad range of GIS skills; large organizations may need several GIS analysts, among whom you might find titles such as these:

- GIS technology expert—responsible for the hardware and network operations of the GIS
- GIS software expert—responsible for application programming
- GIS database analyst—responsible for administration of the GIS database
- GIS primary users
- Professional GIS users—support GIS project studies, data maintenance, and commercial map production
- Desktop GIS specialists—support general query and analysis studies

GIS end users

It is helpful to think of the end users of your GIS as an important staff component because they affect the design and use of the GIS. They may include the following:

- Business experts—key employees with intimate knowledge of the processes your GIS is attempting to improve; they provide the GIS manager with major input regarding design and management of the GIS.
- Customers—clients of the GIS who are served by it, including business users requiring customized GIS information products to support their specific business needs, as well as the more casual Internet and intranet map server users, people accessing basic map products invoked by wizards or web browsers.

Once the GIS is implemented, the management group merits special attention as part of the end-user category. The fundamental purpose of a GIS is to provide to management new or improved information for decision-making purposes. At this point, the managers actually become customers, whereas up until now they have been involved in helping you pass through the hurdles of the organizational bureaucracy. The managers could be the project sponsors, a member of executive management who will be using the application (or its output), or a management representative who can serve as the conduit delivering management's requirements. While not the core GIS staff, they are nonetheless vital to GIS success.

System administration staff

A large organization needs a system administration staff to play an important role in the day-to-day operation of the computer systems supporting the GIS. Staff members may include the following:

- The network administrator—responsible for maintaining the enterprise network
- The enterprise database administrator—responsible for the administration of all databases that interact within the organization
- The hardware technicians—responsible for the day-to-day operation of computer hardware within the organization, including maintenance and repair

Staff placement

Once you've identified the GIS staff required, you must decide where the positions fit into your organizational structure, a decision that will affect the role and visibility of the GIS department. There are four main levels in which GIS staffs are typically placed:

1. Within an existing operational department: In this scenario, staff is tied to a specific need and budget. It is difficult for staff at this level to serve multiple departments.

2. In a GIS services group: This group serves multiple projects but still has the autonomy and visibility of a stand-alone group.
3. At an executive level: This placement signals a high commitment from management. Staff members have high visibility and the authority to help coordinate the GIS project. The downside is when GIS staff members at the executive level become isolated, fostering the perception among the rank-and-file of being out of touch with critical stakeholders.
4. In a separate support department: The more old-fashioned notion of IT would place any new information system staff in a centralized computing services department. This is the "systems" group in many organizations.

GIS job description examples

Clear job descriptions are essential before hiring begins. Outlining GIS roles provides you and your prospective employees with a common understanding about the position and its requirements. Job titles and descriptions will also come into play during job evaluation and performance reviews. Most GIS job descriptions of the same titles are similar; however, they do vary depending on the software used within an organization, size of the system, specific job responsibilities required, and type of agency or company seeking to fill the position.

If your organization is new to GIS, the human resources department will not have suitable job descriptions for the type of GIS personnel you need, so you'll have to write them yourself. Refer to the sample GIS job descriptions here and be sure to design your own just as realistically. Obviously, you want the best person for the job, but if you require that your digitizing technicians have master's degrees, you may not find a digitizing technician.

GIS manager

The GIS manager provides on-site management and direction of services to develop, install, integrate, and maintain an agency-wide standard GIS platform. This position will be responsible for developing, implementing, and maintaining special-purpose applications consistent with the agency's mission and business objectives. Requirements: Applicants must have experience in project-design and work-plan development, database system and application design, and maintenance and administration of a large Oracle SDE database. Applicants must have a bachelor's or master's degree in geography, planning, or a related field, and three to five years professional experience implementing in-depth, complex, GIS solutions involving RDBMS and front-end application development. Applicants must also have knowledge of and the ability to apply emerging information and GIS technologies (particularly Internet and mobile technologies); experience in project management; and excellent interpersonal, organizational, and leadership skills. Experience with object-oriented methods and techniques is a plus.

Enterprise systems administrator

The enterprise systems administrator will provide user support, resolve UNIX- and Windows-related problems, perform systems administrative functions on networked servers, and configure new Windows workstations. Requirements include a bachelor's degree in computer science or other related college degree with experience in computer systems, or three or more years working experience in systems administration functions in a client-server environment. Applicants must be familiar with multiple UNIX platforms and have strong skills in both UNIX and Windows commands and utilities. Good problem-solving skills and the ability to work in a team environment are mandatory.

GIS application programmer

The GIS programmer will design, code, and maintain in-house GIS software for custom applications. This position requires the ability to interpret user needs into useful applications. Applicants should have a bachelor's degree or higher in computer science, geography, or related earth sciences. All applicants must have two years or more of programming experience with one of the following: Microsoft Visual Studio, Python, C++, or GIS vendor-specific programming languages. Previous experience with programming GIS applications and knowledge of GIS search engines are a plus.

GIS database analyst

The GIS database analyst is responsible for the creation of spatially enabled database models for an enterprise GIS. Typical tasks include setup, maintenance, and tuning of RDBMS and spatial data, as well as development of an enterprise-wide GIS database. This position is also responsible for building application frameworks based on Microsoft Visual Studio, Microsoft .Net, and Java approaches. This position requires a bachelor's degree in computer science or geography. Applicants should have a strong theoretical GIS and database design background. Preference will be given to applicants with experience in modeling techniques used by the enterprise.

GIS analyst

The GIS analyst is responsible for the development and delivery of GIS information products, data, and services. Responsibilities include database construction and maintenance using current enterprise GIS software, data collection and reformatting, assisting in designing and monitoring programs and procedures for users, programming system enhancements, customizing software, performing spatial analysis for special projects, and performing QA/QC activities. Applicants must have a bachelor's degree in geography, computer science and planning, engineering, or a related field, or an equivalent combination of education and experience; masters in GIS preferred. Applicants should have one to two years of experience with GIS products and technologies, especially those currently used by the enterprise. Applicants should have experience with various spatial-analysis technologies. Knowledge of current enterprise spatial server engines is a plus.

GIS technician/Cartographer

Responsibilities of the GIS technician or cartographer include all aspects of topographic and map production using custom software applications: compiling data from various source materials, generating grids and graticules, portraying relief, creating map surround elements, digital cartographic editing, placing text, performing color separation, performing quality assurance, creating symbols, and testing cartographic software. Applicants should have strong verbal and written communication skills, and have a bachelor's or master's degree (depending on position level) in geography, cartography, GIS, or a related field. Applicants should have experience or coursework in GIS software and macrolanguages, Visual Studio, Python, or graphic drawing packages. Applicants should be familiar with remote-sensing and satellite imagery interpretation. Applicants should provide a digital or hard-copy cartographic portfolio for evaluation.

Training

Consider how the people within your organization will use the GIS before developing a training program. GIS staff and GIS end users will require different types of training.

Core staff training

The core GIS staff members comprise the cornerstone of your efforts. They are responsible for creating, maintaining, and operating both the data and the system infrastructure. They will require up-front and ongoing training to keep them current on new techniques and methods. (You may need to consider training for system administration staff as well.)

The training program developed for the GIS staff could involve courses in database management, application programming, hardware functionality, or even geostatistical analysis, depending on their relative and collective skill sets. The training your staff receives should complement their job responsibilities. These are the people responsible for maintaining a product for your users. A well-trained staff is crucial for the continual success of a GIS.

End-user training

The training required for GIS end users is understandably much less involved than that for core GIS staff members. Often a single interface or web application is all that an end user will ever see, meaning the application can be taught in as little as minutes.

Many vendors provide training courses related to their own software, and some even cover basic GIS theory and applications as necessary. Self-study workbooks that include software to practice with provide another flexible learning alternative. GIS is inherently a multidisciplinary endeavor, so training in other areas beyond the actual software continues to play a major role.

Manager training

Even if, ideally, the person hired as GIS manager has technical competency in GIS, these skills must be continually updated if the manager hopes to give meaningful direction to his or her staff. Also, of course, GIS managers must demonstrate effective management skills—or work to acquire them. Courses designed to help in specific areas such as general management skills, project management, strategic management, and total quality management can all be helpful.

Training delivery

Training can be delivered in many flexible forms these days, thanks to the web and relatively cheap air travel. Face-to-face classroom courses are available from vendors or educational establishments, either at their premises or on-site with you. Web-based training, distance learning, and self-study workbooks/e-books are all options for training in GIS and related areas. Whatever method is used, sufficient time and resources for training and related activities (travel, preparation of assessment, follow-up reading) must be provided.

Benchmark testing

A benchmark test is a comparative evaluation of different systems in a controlled environment. The test is used to determine which system can handle the anticipated workload in the most cost-effective manner.

Benchmark testing is appropriate if you are planning to procure a large-scale system. It may cost a vendor about $40,000 to put on a benchmark test. If the system you're going to buy will cost $10,000, no vendor is going to conduct a benchmark test for you. The system is too small to make benchmarking cost-effective. That is why such tests are generally conducted only for large system acquisitions, and they are getting harder to recommend to vendors as system prices fall.

A benchmark test is not simply a demonstration of what a system can do. The objectives are to find out whether each system under test can perform the required tasks and to determine each system's relative performance with respect to these tasks. A benchmark test should be designed to verify that the system can perform the functions necessary to create the information products in a timely manner. It should evaluate whether each system can handle the data required, produce the information products needed, and carry out the core functionality. The focus of the tests should be functions that you will use most often. The test is also used for assessing the price of the proposed system compared to its performance.

What you provide vendors

You should provide vendors with information on functional requirements and throughput capacity. Vendors who want to participate in the benchmark testing will then respond with details of a proposed system configuration and the cost of system acquisition and maintenance. The system configuration details would include the type and capacity of the equipment, storage devices, network capacities, and the number of input and product generation workstations. Such details might also be accompanied by descriptions of system capabilities, support services, company characteristics, and contractual commitments in any prebenchmark proposals.

You already gathered information for the functional requirements you need to pass along—you did so during the planning process, when you considered the following aspects of the anticipated workload:

- The information products created by the system in each of the first five years
- The relative importance of each information product to the daily work of the organization (the assigned priority) and the wait tolerance of each information product
- The system functions required to create each information product
- The data required to create the information products
- The yearly volume of each data type required to create the information products

During the conceptual system design for technology, you summarized and classified the functions required by the GIS in a function utilization table and graph. This provides you with the following details:

- The total set of system functions required
- The frequency of anticipated use of those functions
- The relative importance of those functions

Having determined the functionality required, and thereby the anticipated workload, you can now establish the throughput capacity of the system. Throughput capacity is the amount of work that the GIS must be able to carry out over a period of time, which is a function of the combined capabilities of the system software, hardware, and network bandwidth.

Testing guidelines

A benchmark test is a compromise between exhaustive and inadequate testing. You should aim to keep the test as compact as possible while still producing reliable results. To minimize the testing effort placed on the vendors, while still allowing yourself to gather the information required, you can adopt the following guidelines:

- Send approximately 85 percent of the test data and all of the test questions to the vendors two months in advance of the test. This gives the vendors ample time to create the required databases and set up their systems to perform at optimum levels during the test.
- Make sure that each vendor receives the same data and test questions. This uniformity permits the testing of all vendors on an equal basis. It also permits you to create testing scenarios with known answers.
- Provide approximately 15 percent of the test data on the first day of the benchmark test to permit real-time observation of data entry and data updating.
- Choose a set of information products that is representative of your organization's needs, including some that require rapid response, some that are mission critical, and some that are high-capacity business support applications.
- Carefully choose which information products to test to ensure that the proposed system has the capacity to create the other information products required by your organization.
- Test most thoroughly those frequently used functions required for the highest-priority information products. Less-used functions or lower-priority functions should be tested the least, but all functions should be tested at least once. Any insufficiently tested functions can be tested in a separate section of the test specifically designed for that purpose.

Logistics of testing

You need to address other questions concerning the logistics of testing. These include the following sections.

Where should the test be conducted?

It is usually preferable to hold the benchmark test at a site chosen by the vendor. They may choose their own company headquarters. However, if the new system will be embedded within an existing enterprise network, it is better for the client to provide an isolated network that emulates the multiple data transmission speeds encountered in their network with wide area connections. If the client provides the testing site, the vendor should supply the proposed hardware at the client's site using the vendor-recommended protocols. If it is not possible to establish the network test at the client's site, the network may be simulated at the vendor's site with provision made to closely monitor network traffic volumes.

When should the test be carried out?

The vendors should select the test dates. Once the testing dates are chosen by all the vendors, send the questions and materials to the vendors by courier at appropriate intervals to ensure that each vendor has the same amount of time between when they receive the materials and their scheduled testing date.

Who should manage the testing?

A benchmark test team should manage the process and evaluate the results. This team should be composed of key client personnel, such as users, technical-support staff, consultants, and managers. It is important for the users to have a sense of ownership in the decision. It is also wise to include technical support and management staff. They will be crucial allies during implementation and in providing ongoing support. These different perspectives will help make the selection decision a better one.

Who should monitor the tests?

A subset of the management team should monitor the actual tests. Ideally, this monitoring team should consist of a minimum of two and a maximum of four people, including technical staff with previous benchmark test experience or one or two experienced GIS consultants, plus one or two staff from your organization who have firsthand knowledge of the GIS analysis required. The same monitoring team should preside over each benchmark test in the procurement.

Who pays for the testing?

Current practice is to share benchmark test costs between the client and the vendors. The client typically underwrites the cost of test preparation, materials, provision of an isolated network for testing, test monitoring, result analysis, and reporting. The vendors accept costs of database creation, installation of hardware on the isolated network (if required), and performing system tests.

Evaluation and scoring

Benchmark testing is conducted to ensure that the proposed system can perform the functions you need to generate your information products. Without testing a proposed system adequately, you risk making a purchase with little value to your organization. Proper evaluation is part of adequate testing.

Your evaluation of the functional requirements should answer the following questions:

- Can the systems being tested perform the functions specified in the request for proposal (RFP)?
- What is the relative performance of different systems on a function-by-function basis?
- What is the effect of the proposed system configuration and network utilization design on the ability of the system to generate the required information products in a timely manner?

The functions you wish to evaluate should be identified in each section of any benchmark test guidelines provided to the vendors. Each information product has performance requirements (e.g., wait tolerance) specified.

A benchmark test team, together with anyone involved in preparation of data and answers for the benchmark test, should evaluate the functionality. (The monitoring team is an important part of the benchmark test team.) To evaluate functionality, the benchmark test team needs the following:

- All observations made by the monitoring team during the benchmark tests
- The indicators of system performance by function measured during the benchmark tests
- The results of the verification performed after the benchmark tests

Each function should be allocated a score using the scoring system from shown in figure B.1. This scoring system should be provided to vendors in the RFP. When the test is complete, the benchmark test team should discuss each function to arrive at a unanimous decision for the final scores.

Now you should assess how the production of information products will be affected by the functional capability. To do this, use your list of information products ranked by priority. Assign a 0–9 function score to each function invoked to make each product, determine the total and average score for each product, and determine the worst score for any of the functions required to make each product.

These results will be sufficient where there is a substantial difference in functional capabilities between the systems under evaluation in the benchmark test. It should be clear if one or more of the systems includes restricted functionality that prohibits the making of high-priority, frequently used products.

Where differences between systems are less pronounced, you'll need to produce a graph like the one shown in figure B.2 to help in your decision making. Plot the highest function score obtained for each information product on the vertical axis against the priority ranking for each information product (on the horizontal axis). Plot a separate curve for each system you have tested.

Score	Function appraisal	Criteria
0	Outstanding	All of the qualities in appraisal totally integrated into an operational system—the best in the industry.
1	Excellent	Elegant, well-thought-out solution for individual functions—very fast and user friendly.
2	Very good	Fast and user friendly.
3	Good	Adequate and fast or user friendly.
4	Satisfactory	Adequate.
5	Functional	Function can be performed. Needs minor improvement for speed or ease of use.
6	Functional with limitations	Function can be performed. Needs substantial improvements for speed or ease of use.
7	Partial only	New software development required for part for function.
8	Absent or not demonstrated	New software development required.
9	Absent and constrained	Impossible or very difficult to implement without major system modification.

Figure B.1 **Criteria for system evaluation.**

If there are no overlaps of the separate curves for each system on your graph, then the preferred system on functional grounds will be the one appearing highest on the graph (as on the left in figure B.2). If the curves on your graph overlap (as on the right in figure B.2), you must make a judgment depending on the relative importance of high-priority versus low-priority products.

Throughput capacity

Again, throughput capacity is the amount of work that your system must be able to carry out over a period of time. To evaluate throughput capacity you should determine system performance for data input, information product generation, and network capacity utilization.

Data input

Calculate personnel time required for data input. When setting up your benchmark tests make sure you can extrapolate personnel time for each dataset planned for input in each year of the five-year planning horizon. Determine data storage requirements; estimate these in terabytes, gigabytes, megabytes, or kilobytes for each dataset as appropriate.

Information product generation

Estimate the data volume by data function for each information product. Use the master input data list and the information product descriptions to generate this estimate. Use the measures of system performance by function and data volume collected during the benchmark test to calculate personnel time, CPU time, network capacity utilization, and plotting time to generate each information product (if plotting is needed).

Compare these results with the results for the same variables generated during the benchmark test. After the benchmark test, you should be able to estimate the following, by year:

- The total personnel time required to input data
- The core performance metrics required to input data and create information products
- Storage requirements of the system

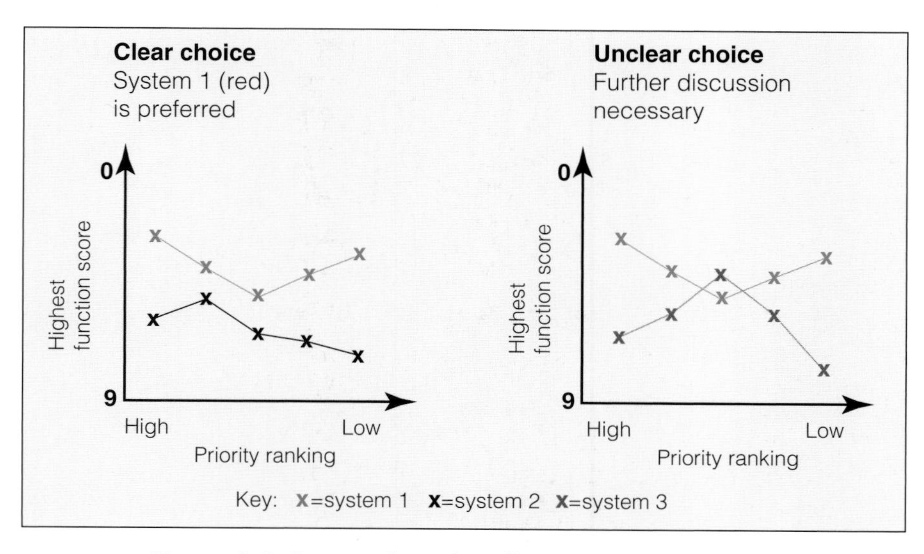

Figure B.2 **System functionality comparison graph.**

- The network capacity utilization to create the required information products at the specified wait tolerances
- The plotter hours required for information product generation

Compare the results of personnel time used for daily input and information product generation with personnel availability. Convert the results to operational costs by year using hourly rates, so you can assess the overall use costs for each system.

The core utilization figures will require adjustment if the core type used by the vendor during the benchmark test is different than the core metrics proposed for acquisition by your organization. When necessary, you can use the relative performance ratios provided by computer manufacturers. In general, advise the vendor to use the proposed core type for the benchmark test. Compare the results of the system resource use calculations with the system configuration proposed by the vendor. This will allow you to determine if the proposed configuration can handle the workload.

Network capacity utilization

When enterprise networks are needed for information product generation and tests have been conducted on an isolated controlled network, a quantitative evaluation of network capacity utilization is possible.

The benchmark test results allow you to calculate the effect on the network caused by use of the GIS. The benchmark test gives the percentage of network capacity utilization by information product for single and multiuser conditions. These numbers can be extrapolated to estimate, by year, the network capacity utilization associated with the generation of the proposed information products. Estimates will reflect the hardware and protocol options and data search engine optimization design recommended by the vendor. You should compare these estimates with the client network administration numbers on network percent capacity availability in the same time period.

The measures to be tested should be specified in the benchmark test guidelines and sent to vendors.

Network design planning factors

Client platform	Data per display		Traffic per display		Kb per second traffic per user	
	Kb per display	Adjusted Kb per display	Kb per display	Mb per display	6 displays per minute	10 displays per minute
File server client	1,000	5,000	50,000	50.000	5,000	8,333
Geodatabase client	1,000	500	5,000	5.000	500	833
Terminal client						
Vector only data source	100	28	280	0.28	28	47
Imagery data layer	100	100	1,000	1	100	167
Web browser client						
Light complexity	100	100	1,000	1	100	167
Medium complexity	200	200	2,000	2	100	333
Web GIS desktop client						
Light complexity	200	200	2,000	2	200	333
Medium complexity	400	400	4,000	4	400	667

Figure C.1 **Network load factors.**

A system design performance factor is derived for five primary ArcGIS architectures, starting with average data requirements per query. Each of these analyses follows.

File server client

A file server client architecture represents a standard GIS desktop client accessing data from a file server data source. Data required to support display is 1,000 KB per query. Traffic between the client and server connection is 5,000 KB per query (file must be transferred to client to support query of data extent required for display). Converting data to traffic yields 50,000 Kb or 50 Mb per display (8 Kb per KB with 2 Kb overhead per traffic packet). Traffic per user depends on productivity, 5,000 Kbps with 6 displays per minute and 8,333 Kbps with 10 displays per minute. Typical GIS power users generate an average of 10 displays per minute.

DBMS client

A DBMS client architecture represents a standard GIS desktop client accessing data from a DBMS data source. Data required to support display is 1,000 KB per query. Data is compressed in the geodatabase, reducing data-transfer requirements to 500 KB per query. Converting data to traffic yields 5,000 Kb traffic per query (8 Kb per KB with 2 Kb overhead per traffic packet). Traffic per user depends on productivity: 5,000 Kbps with 6 displays per minute and 8,333 Kbps with 10 displays per minute. Typical GIS power users generate an average of 10 displays per minute.

Caution: The two client/server architectures above are normally supported over an Ethernet LAN environment. Only one transmission can be supported over a shared LAN segment at a time; when multiple transmissions occur at the same time with current switch technology, transmissions are cached on the switch and fed sequentially across the network. Too many users on a shared segment can result in performance delays. For this reason, maximum traffic on shared Ethernet segments is typically reached by about 50 percent of the total available bandwidth (due to probability of transmission delays at higher utilization rates).

Terminal client

A terminal-client architecture represents terminal client access to standard GIS desktop software executed on a Windows Terminal Server. Data required to support display is 100 KB per query (display pixels only). Traffic between the client and server connection is 28 KB per query (average 75 percent compression with Citrix ICA protocol) for vector-only data sources. Converting data to traffic yields 280 Kb traffic per query (8 Kb per KB with 2 Kb overhead per traffic packet). Traffic per user depends on productivity, 5,000 Kbps with 6 displays per minute and 8,333 Kbps with 10 displays per minute. Typical GIS power users generate an average of 10 displays per minute.

Terminal client display traffic does not compress as well when the data source includes an imagery layer. Traffic between the client and server connection is 100 KB per query, which converts to 1,000 Kb traffic per query. Traffic per user is 100 Kbps for 6 displays per minute and 167 Kbps for 10 displays per minute productivity.

Web browser client

Web browser client architecture represents browser client access to a standard web map service. Data required to support a light complexity display is typically 100 KB per query (typical size of image data transfer generated by Web mapping service). Traffic between the client and server connection is 100 KB per query (no additional compression).

Converting data to traffic yields 1,000 Kb traffic per query (8 Kb per KB with 2 Kb overhead per traffic packet). Traffic per user depends on productivity, 100 Kbps with 6 displays per minute and 167 Kbps with 10 displays per minute. Typical GIS web clients generate an average of 6 displays per minute. Note: If peak map request rates are given, 1,000 Kbpq would be the more accurate network design factor for assessment purposes.

Data required to support a medium-complexity display is typically 200 KB per query. Traffic between the client and server connection is 200 KB (or 1,000 Kb) per query. This equates to 200 Kbps with 6 displays per minute and 333 Kbps with 10 displays per minute.

Web GIS client

Web GIS client architecture represents GIS desktop client access to a standard web mapping service. Data required to support a light complexity display is 200 KB per query (based on typical user-selected display resolution—pixels per display determines traffic requirements). Traffic between the client and server connection is 200 KB per query (no additional compression). Converting data to traffic yields 2,000 Kb traffic per query (8 Kb per KB with 2 Kb overhead per traffic packet). Traffic per user depends on productivity, 100 Kbps with 6 displays per minute and 167 Kbps with 10 displays per minute. Typical GIS web clients generate an average of 6 displays per minute. Note: If peak map request rates are given, 2,000 Kbpq would be the more accurate network design factor for design purposes.

Data required to support a medium-complexity display is typically 200 KB per query. Traffic between the client and server connection is 200 KB (or 1,000 Kb) per query. This equates to 200 Kbps with 6 displays per minute and 333 Kbps with 10 displays per minute.

Caution: The three architectures above (terminal client, web browser client, and web GIS client) are normally supported over WAN environments. Only one transmission can be supported over a shared WAN segment at a time; when two transmissions occur at the same time, transmissions are cached at the router and fed sequentially onto the WAN link. Transmission delays will occur as a result of the cache time (time waiting to get on the WAN). For this reason, optimum performance on shared WAN segments is typically reached with less than 50 percent of the total available bandwidth (due to probability of delays at higher utilization rates).

Acknowledgments

The author would like to express appreciation to Dave Peters of Esri for providing these network planning factors and the network traffic transport time and performance per CPT tables in chapter 9 and appendixes D and E. For more detailed information, please refer to his System Design Strategies wiki site referenced in the "Further Reading" section of this book.

Appendix D

Custom workflows

This section expands on the workflow performance target selection discussed in chapter 9. Standard workflow performance targets (processing loads expressed as seconds of component usage) are based on a reasonably conservative information product design specification. Custom workflow performance targets modify baseline performance to reflect more specific information product specifications by adjusting general results to more accurate workflow loads. The system design analysis is only as good as the user workflow loads input: more accurate input provides more accurate results.

Custom workflows

The factors needed to design the custom workflows can be identified and made part of the workflow name, and so provide a workflow recipe for the redesign. This shorthand prescription allows the custom workflow service times to be computed and the recipe carried through as a software design specification.

Processing and network traffic loads are unique to each user workflow. Custom workflow loads can be generated by modifying standard workflow baseline loads according to selected adjustments. Figure D.1 shows how the standard workflow baseline loads can be modified by the selected deployment parameters.

Standard workflow baseline processing loads are predefined for the most common software technology patterns, derived from software vendor performance validation benchmark testing. These baseline loads are modified by map document selection, display complexity, percentage of data that resides in a cache, display resolution, and output format—important performance parameters required for specific information product specifications. The final custom workflow processing loads are a product of

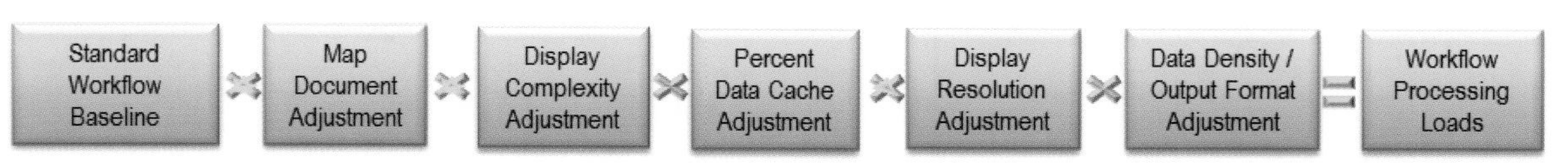

Figure D.1 **Custom workflow load generation.**

Workflow	Software Component Service Times							Software Service Time
		Arc12 baseline				SRint06/cor 45.0		
	Client	Design Model Metrics				Database		
	Traffic	Client		Web	SOC	Data	Data	
Standard Workflows =========	Mbpd	Client	Citrix	Web2	SOC	SDE/MDS	DBMS	Total
GIS for Desktop =========	Mbpd	Client	Citrix	Web2	SOC	SDE/MDS	DBMS	Total
AGD101 wkstn MXD 100%Dyn Med 10x7 Feature	10.000	0.338				0.042	0.042	0.422
AGD101 Citrix MXD R 100%Dyn Med 10x7 ICA	1.000	0.036	0.338			0.042	0.042	0.458
ArcGIS for Server =========	Mbpd	Client	Citrix	Web2	SOC	SDE/MDS	DBMS	Total
AGS101 REST MSD R 100%Dyn Med 10x7 JPEG	2.000	0.036		0.012	0.116	0.029	0.029	0.220
AGS Full MapCache Service	0.500	0.036		0.009	0.009			0.053
GIS for Mobile =========	Mbpd	Client	Citrix	Web2	SOC	SDE/MDS	DBMS	Total
AGS101 Mobile Client		0.166						0.166
AGS101 Mobile Sychronization Service	0.031	0.036		0.019	0.015	0.001	0.001	0.073

Figure D.2 **Standard workflow performance baseline component processing loads.**

the selected software technology and performance adjustments for each user workflow.

Standard workflow baselines

The starting point is selecting the appropriate standard software technology baseline. The selected standard workflow service times are used as the starting point to generate custom workflows performance targets. Standard workflows have a standard map document (MXD), medium display complexity, are 100 percent dynamic (full map cache service is 100 percent cache map), 1024 × 768 display resolution, and image raster density with JPEG output format.

Performance baselines are established for the most common workflow patterns. Service times shown in figure D.2 are relative to a selected 2012 baseline platform (Intel Xeon E5-2643 processor), where the published SPEC benchmark baseline per core is equal to 45.

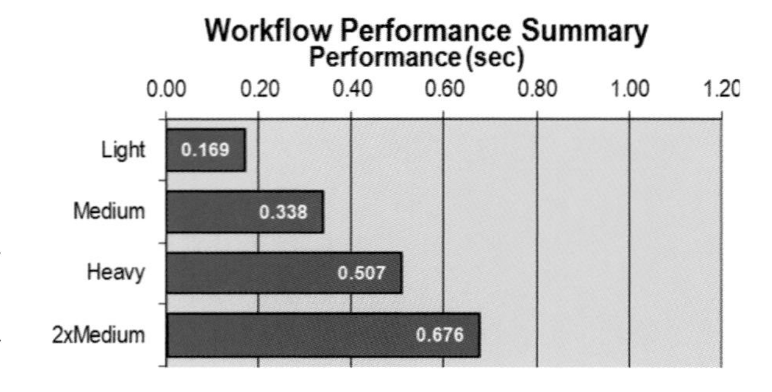

Figure D.3 **Map display rendering time.**

Optimized map document

The first adjustment shown in figure D.1 accounts for the selected map document publishing format. ArcGIS 9.3.1 introduced an optimized map service description (MSD) that uses a new graphics rendering engine. The new rendering engine improves display quality and reduced ArcGIS Server SOC processing time by 30 to 70 percent.

For planning purposes, reduce SOC baseline processing time by 50 percent when publishing map services using an MSD map document (this applies only to published services).

Display complexity

All maps are not created equal. More layers, features, and sophisticated geoprocessing outputs increase map display complexity. Figure D.3 shows measured rendering performance for light, medium, and heavy map displays. Most of the desktop rendering time is client processing. Network traffic, queue time, and latency times are negligible. Map display complexity is a measure of the map document desktop rendering time.

The first step in authoring a web mapping service is to create your information product (MSD). To measure map display complexity, use the desktop application with a local file geodatabase data source and measure the average information product rendering time. Select an average display that you feel will represent your workflow loads. The display rendering time is measured on your workstation, which may perform much slower than the current performance platform baseline. Guidelines for translating your measured display performance to the current platform performance baseline are provided in figure D.4.

The map document rendering time—one way to measure display complexity—can be measured using the Desktop Map Publishing tools.

Map display complexity is related to the map display rendering time. For the baseline platform shown in figure D.4, a workflow with an average display render time of 0.34 sec is considered a medium-complexity display. A light display is rendering in half this time (0.17 sec), and a heavy display takes 50 percent more processing time (0.51 sec). These are workflow performance targets that should be carried forward through software design and deployment. Once deployed, the service time should be

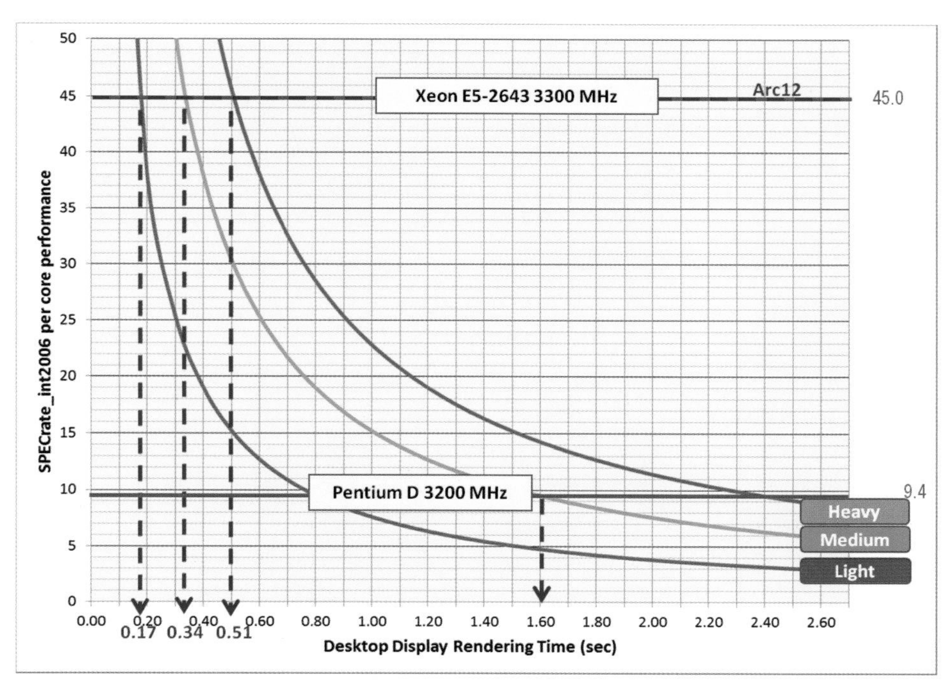

Figure D.4 **Measured display complexity.**

measured to validate the solution is within established design specifications.

The baseline map display service times identified in figure D.2 represent medium display complexity map services. A complexity adjustment factor (light = 0.5, medium = 1, heavy = 1.5) is applied to the baseline GIS Desktop client, SOC, SDE, and DBMS software component service times.

Map cache adjustments

One way to improve map display performance is to preprocess and store the information product in an optimized map cache pyramid. The information product often includes a mix of static basemap reference layers or imagery combined with a map overlay of several rapidly changing business (operational) layers. The basemap layers can be preprocessed and stored in a map cache, reducing the number of layers that must be dynamically rendered for the map service. Reducing the number of dynamic layers reduces the display rendering time.

The map author can measure desktop rendering time with and without the cached basemap layers to estimate the percentage of display in the map cache. GIS Desktop client, SOC, SDE, and DBMS service times are reduced by the percentage of map display that is cached.

A dynamic map service mashup with a map cache service layer will require additional display traffic. The initial map display traffic will include the dynamic service image and the map cache tiles. The initial map cache tile download traffic can be 2 to 4 megabits per display in addition to the dynamic map service image. Once the map tiles are available in the client browser cache, repeat display requests for the same tiles will no longer be downloaded from the server. Client workflows that work in the same area will soon have the required tiles in local cache, and additional tile cache traffic will be minimal. Clients that move to different areas with each display will have increased traffic loads due to the tile cache transfers. For planning purposes, increase display traffic by about 0.5 Mbpd over the dynamic service traffic to accommodate the tile traffic loads (assumes about 75 to 85 percent of images are being pulled from local cache tiles).

Fully cached web mapping services provide the best user performance. A minimum level of server processing per display transaction is required to deliver the cached

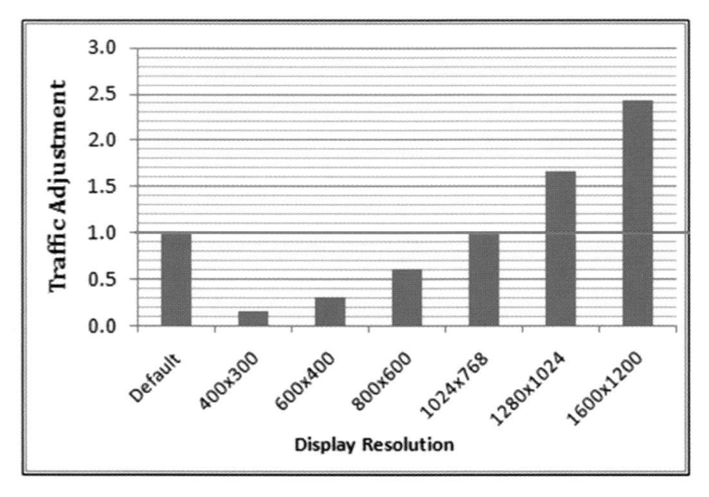

Figure D.5
Display resolution network traffic adjustments.

tiles (web = 0.01 sec, SOC = 0.01 sec). Additional web loads may be required when using a condensed map cache due to the file handler processing loads (web = 0.02 sec/display).

Display resolution

Client map display resolution can affect display performance and web service display traffic. The primary effect is on web traffic, since image display resolution determines the number of pixels that must be delivered to the client. Higher display resolution means more pixels, and more pixels translate to more display traffic. The effect on network traffic is linear in nature, based on the total number of pixels in the display, as shown in figure D.5.

Display service times can also be affected by resolution, because display resolution can affect map scale, and larger map scales can include more features and functions. More features will increase required map processing time (service time). Figure D.6 identifies performance adjustments that should be made to accommodate specific display resolutions. Performance adjustments are made to the SOC, SDE, and DBMS software component service times.

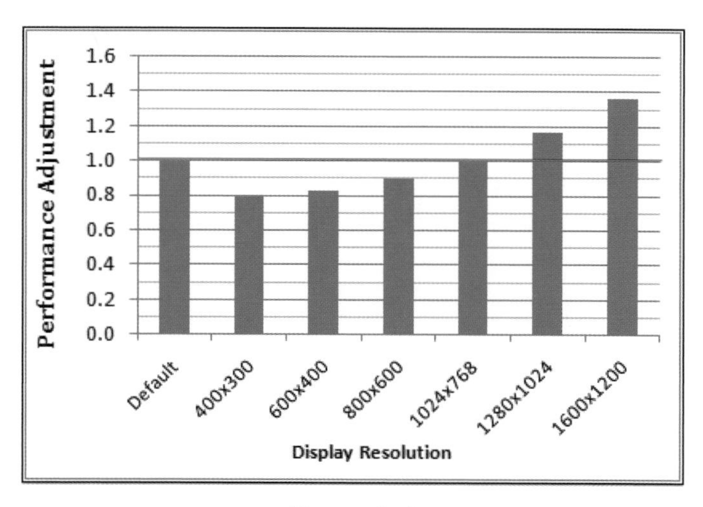

Figure D.6
Display resolution performance adjustments.

Display resolution adjustment factors for both traffic and application processing loads (performance adjustments) are provided in figure D.7. As you can see, most of the resolution adjustment effect is on network traffic.

Display Resolution	Traffic Factor		App (SOC,SDE,DB)
Default	1.0	10x7	1.00
400x300	0.15	4x3	0.79
600x400	0.31	6x4	0.83
800x600	0.61	8x6	0.90
1024x768	**1.00**	**10x7**	**1.00**
1280x1024	1.67	12x10	1.17
1600x1200	2.44	16x12	1.36

Figure D.7
Display resolution adjustment factors.

Display density and output format

Density is a term used to identify the color variation between image service display pixels, an important consideration for web mapping. Images generated from a vector-only data source compress much better than images generated from a data source that includes orthophotographs or other satellite imagery. Since imagery is a picture of Earth's surface, each pixel in the display can vary slightly from those around them. This variation limits the compression performance with different image output formats, as shown in figure D.8.

JPEG format provides the best output compression for imagery data sources. PNG8 format provides the best output compression for vector-only data sources. PNG24 and PNG32 images are required for higher color depth; however, this additional quality comes with a traffic penalty, particularly if the data source includes digital imagery layers.

Software processing loads also vary between the different output formats. The higher quality PNG output formats require additional processing loads. Figure D.9 provides an overview of performance adjustments impacting SOC and web display service times.

Figure D.10 provides a summary of the display output performance adjustment factors. Performance factors are different for vector-only and imagery data sources.

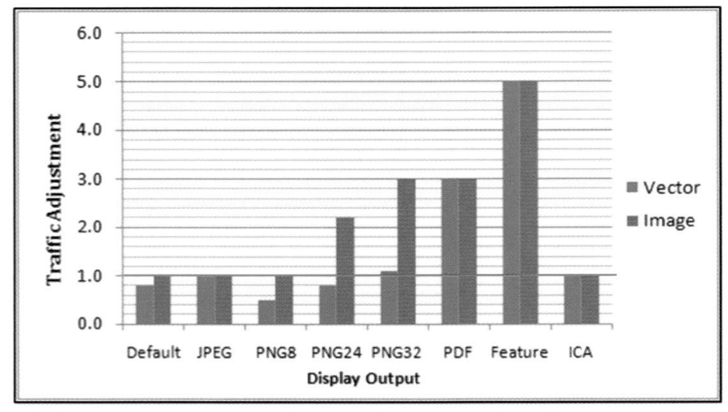

Figure D.8 **Display output traffic adjustment.**

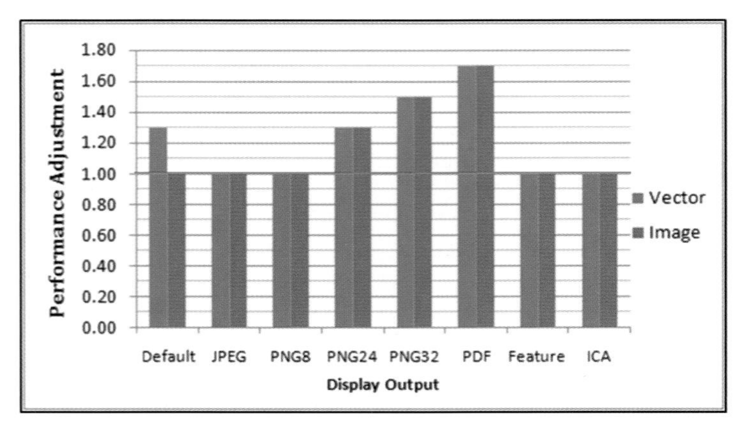

Figure D.9 **Display output performance adjustment (SOC and web service times).**

Network traffic is affected more than server processing loads. The default image output for vector data sources is PNG24 (adjustment factor of 1.3 times the standard ArcGIS Server workflows, which include raster imagery by default).

Output	Traffic		App (Web,SOC)	
	Vector	Image	Vector	Image
Default	0.8	1	1.3	1
JPEG	1	1	1	1
PNG8	0.5	1	1	1
PNG24	0.8	2.2	1.3	1.3
PNG32	1.1	3	1.5	1.5
PDF	3	3	1.7	1.7
Feature	5	5	1	1
ICA	1	1	1	1

Figure D.10
Display output adjustment factors.

Workflow recipe

The software technology selection and established performance patterns identified earlier establish custom workflow performance targets used to complete the system architecture design. The assumptions used in establishing these performance targets now become development specifications that must be carried through the software design and deployment process. If these workflow design specifications are not followed during service development and deployment, the workflow traffic and processing loads used during planning to generate required hardware and network specifications are no longer valid.

Figure D.11 introduces a custom workflow nomenclature used as metadata to describe the software technology selection and performance parameter adjustments used to generate the workflow performance targets. Figure D.12 defines the software abbreviations used in the workflow recipes and charts.

The workflow name is generated from the technology selections made in establishing the custom workflow performance targets. The workflow captured by figure D.11 is an ArcGIS 10.1 for Server REST software technology pattern (abbreviations are shown in figure D.12), deployed with an optimized map document (MSD), light complexity map document, 100 percent dynamic (no preprocessed map cache), 1024 × 768 output display resolution, and vector-only data source with a JPEG output format. An additional 0.5 Mbpd can be

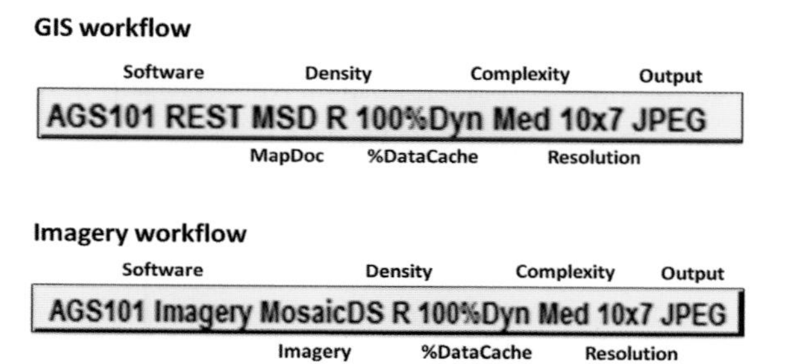

Figure D.11
Example of a custom workflow recipe.

Software technology	
AGD101 wkstn	ArcGIS 10.1 for Desktop workstation
AGD101 Citrix	ArcGIS 10.1 for Desktop WTS Citrix
AGS101 REST	ArcGIS 10.1 for Server REST service
AGS101 WMS	ArcGIS 10.1 for Server WMS service
AGS101 SOAP	ArcGIS 10.1 for Server SOAP service
AGS101 Imagery	ArcGIS 10.1 for Server Image service
AGD10 wkstn	ArcGIS Desktop 10 workstation
AGD10 Citrix	ArcGIS Desktop 10 WTS Citrix
AGS10 REST	ArcGIS Server 10 REST service
AGS10 WMS	ArcGIS Server 10 WMS service
AGS10 ADF	ArcGIS Server 10 ADF service
AGS10 SOAP	ArcGIS Server 10 SOAP service
AGS10 Imagery	ArcGIS Server 10 Image service
AGD931 wkstn	ArcGIS Desktop 9.3.1 workstation
AGD931 Citrix	ArcGIS Desktop 9.3.1 WTS Citrix
AGS931 ADF	ArcGIS Server 9.3.1 ADF service
AGS931 SOAP	ArcGIS Server 9.3.1 SOAP service
AGS931 REST	ArcGIS Server 9.3.1 REST service
AGS931 WMS	ArcGIS Server 9.3.1 WMS service
AGS Full MapCache	ArcGIS Server Full MapCache service

Figure D.12 **Software technology patterns and abbreviations.**

added to an account for additional traffic due to display mashup with an online cached basemap data source. A similar recipe is provided for imagery workflows, with choice of imagery dataset (Mosaic or Raster) replacing the MapDoc selection. You will use this recipe to generate custom performance targets for the City of Rome web maps workflow.

Figure D.13 shows how to use the workflow recipe to calculate custom workflow performance targets for the system architecture design. To achieve custom workflow performance targets, start with the standard workflow baseline service times, then multiply by the adjustment factors for each software performance parameter (a factor of 1 is used when there is no adjustment required).

The custom workflow generation values used in this section were developed from vendor benchmark testing. Each software release should include testing to validate these performance relationships and update these adjustments. Capacity planning can only be as accurate as our understanding of these performance relationships.

Software development and deployment must follow the same assumptions made during planning if we expect to see the same results. Performance should be tested during design and deployment to validate performance targets are met. Planning sets a foundation for system deployment. The plan must be followed during development and deployment to deliver within the performance budget.

Perf Factor	Workflow	Client traffic	client	Citrix	Web	SOC	SDE/MDS	DB Traffic	DBMS	Total
Adjusted baseline svc times		2.1 Mbpd	0.036		0.015	0.150	0.029	10 Mbpd	0.029	0.259
MapDoc	MSD					0.5				
Complexity	Med	1	1	1		1	1	1	1	
Dynamic	100%	100%	100%	100%		100%	100%	100%	100%	
Resolution	10x7	1.00	1.00	1.00		1.00	1.00		1.00	
Density	V									
Output	PNG24	0.8			1.3	1.3				
+MapCache	+mapcache	0.5								
Base Workflow	AGS101 REST	2 Mbpd	0.036		0.012	0.231	0.029	10 Mbpd	0.029	0.336

Figure D.13 **Custom workflow recipe calculation.**

Appendix E

City of Rome system architecture design alternatives

Chapter 9 provided the basic system architecture design process for City of Rome. City of Rome is planning a GIS enterprise system deployment that would initially span over two years. Several design options were considered before making a final design decision. These design options are:

City of Rome year 1

- Minimum physical platform configuration
  - Alternate option 1: High-availability physical platform configuration
  - Alternate option 2: High-availability virtual server platform configuration

City of Rome year 2

- High-availability virtual server platform configuration
  - Alternate option 1: High-availability virtual server configuration with public web services deployed on the Amazon cloud

This appendix provides the details for the alternate options. You should first read the case study in chapter 9 before reading on.

Year 1 minimum physical server configuration

The City of Rome system architecture design began with a user needs workflow analysis. User location and network connectivity was identified for each user site (figures 9.13 and 9.15). User workflow requirements were identified by department for each deployment phase (year 1, year 2), identifying peak workflow loads (users or requests per hour) for each user workflow at each physical location (figure 9.17). Custom load profiles were identified along with appropriate software service times for each identified user workflow (figure 9.19). Once the user requirements needs were identified, a network suitability analysis was conducted (figure 9.20) to establish required year 1 network bandwidth requirements (figure 9.21). A workflow platform loads analysis (figure 9.23) was then completed to identify selected platform loads. Once the baseline platform loads were identified for each selected platform tier, a platform selection and pricing analysis (figure 9.25) was performed to complete final configuration and pricing.

The complete analysis for the year 1 minimum physical server configuration was detailed in chapter 9. The final configuration included a four-platform tier (WTS, WebIn, WebPub, DBMS). The minimum solution included four Xeon E5-2637 4-core servers. The total hardware cost for the minimum physical server configuration of $55,740 was shown in figure 9.25.

Server Platform tier	Adjusted sec	Percent Rollover	Required Core	Platform Core	Platform Nodes	RAM per Node	Node Cost	Total Cost	Platform Utilization
WTS: WTS Platform Tier	96.1	80%	1.6	4	2	48 GB	$14,385	$28,770	20%
WebIn: Internal Mapping Tier	62.1	80%	1.0	4	2	12 GB	$13,485	$26,970	13%
WebPub: Public Mapping Tier	29.2	80%	0.5	4	2	12 GB	$13,485	$26,970	6%
DBMS: Database Tier	31.3	80%	0.5	4	2	48 GB	$14,385	$28,770	13%
Storage pricing estimates: Volume of data gigabites + 50% (data indexing) x $ per GB at required RAID level (see data storage estimates)									
							Total cost =	$111,480	

Figure E.1 **Rome year 1 high-availability physical server platform selection and pricing.**

City of Rome Year 1	User Environment					Bandwidth Mbps	NW %Cap	Platform Service Times Arc12 4 core (1 chip) Baseline				
Types of Workflows	Peak Concurrent		DPM/TPM		Network		Traffic					
Standard	Users	TPH	per Client	Total	Mbps		Mbpd	Client	WTS	WebIn	WebPub	DBMS
LAN_Local Clients		Services	47 Clients	LAN = 31.4 Mbps		100	31%	Platform Service Times (seconds)				
GISDeskEditor	15		10.00	150	13.750		5.500	0.190				0.021
GISDeskView	31		10.00	310	15.500		3.000	0.096				0.011
WebInternal		4,600	6.00	77	2.172		1.700	0.036		0.104		0.014
BatchAdmin	1		249.33	249						0.157		0.029
WAN_Clients			52 Clients	WAN = 7.2 Mbps		24	30%					
Site 2_Operations			2 Clients	Trafic = 0.7 Mbps		3	24%					
RemoteGISView	2		10.00	20	0.237		0.710	0.036	0.190			0.021
WebInternal		1,000	6.00	17	0.472		1.700	0.036		0.104		0.014
Site 3_Freeberg			30 Clients	Trafic = 3.8 Mbps		12	32%					
RemoteGISView	30		10.00	300	3.550		0.710	0.036	0.190			0.021
WebInternal		500	6.00	8	0.236		1.700	0.036		0.104		0.014
Site 4_Willsberg			20 Clients	Trafic = 2.7 Mbps		12	22%					
RemoteGISView	20		10.00	200	2.367		0.710	0.036	0.190			0.021
WebInternal		700	6.00	12	0.331		1.700	0.036		0.104		0.014
Internet_Clients				Internet = 14.2 Mbps		24	59%					
Public_Web Services				Trafic = 14.2 Mbps		24	59%					
WebPublic		30,000	6.00	500	14.167		1.700	0.036			0.060	0.007
Total Throughput	99	36,800										

Figure E.2 **Rome year 1 virtual server workflow platform loads analysis.**

Year 1 high-availability physical server configuration

The year 1 high-availability physical server configuration will require a minimum of two servers for each platform tier. The Xeon E5-2637 4-core servers and the loads analysis in figure 9.23 can be used to complete the sizing analysis. Figure E.1 completes the analysis adjusting the final calculations to include a minimum of two platform nodes for each tier.

The WTS, WebIn, and WebPub tier will balance the processing load across the available server platforms. When adding an additional server to each tier, platform utilization will be reduced by 50 percent (WTS = 20 percent, WebIn = 13 percent, and WebPub = 6 percent). The DBMS tier will be configured for failover operations and include a backup server that will take up the load if the primary server fails (peak platform utilization for the active server will remain at 13 percent). Total hardware cost for the high-availability server configuration in figure E.1 is $111,480.

Total baseline platform loads								
Client	WTS		WebIn		WebPub		DBMS	
Users	Users	Sec	Users	Sec	Users	Sec	Users	Sec
15.0							15.0	3.2
31.0							31.0	3.4
12.8			12.8	8.0			12.8	1.1
			1.0	39.1			1.0	7.2
2.0	2.0	3.8					2.0	0.4
2.8			2.8	1.7			2.8	0.2
30.0	30.0	57.0					30.0	6.3
1.4			1.4	0.9			1.4	0.1
20.0	20.0	38.0					20.0	4.2
1.9			1.9	1.2			1.9	0.2
83.3					83.3	30.0	83.3	3.5
200.2	52.0	98.8	19.8	50.9	83.3	30.0	201.2	29.8

Figure E.2 (continued)

Year 1 high-availability virtual server configuration

The high-available virtual server configuration will deploy multiple virtual server machines on selected physical host machines. The City of Rome will use the high-capacity Xeon E5-2667 12-core (2-chip) 2900 MHz platforms for their virtual server host machines. The selected host platform will determine the processing performance for the virtual server machines. The per-core performance on the Xeon E5-2667 processors are slower than the Xeon E5-2637 processors we used for the physical platform solutions.

The workflow platform loads identified in figure E.2 can be used as a starting point for completing the virtual server sizing analysis. The slower host processor speeds (SPECrate_int2006 = 41.8 per core) will reduce the BatchAdmin workflow productivity and change the baseline processing loads for the WebIn and DBMS host platforms. The virtual server environment includes some processing overhead (10 percent per virtual core) that must be included in the analysis. The adjusted BatchAdmin service times (WebIn = 0.203, DBMS = 0.038) are adjusted using the slower Xeon E5-2667 processor SPEC per-core baseline. The new BatchAdmin productivity is 249 with the adjusted server loads (WebIn = 39.1 sec, DBMS = 7.2 sec). Adjusted loads shown in figure E.2 can be used to complete the sizing analysis.

The virtual server environment includes the host platform and the virtual server configuration. The host platform determines the virtual server per-core performance. The virtual server machines are configured as 2-core servers. There is a 20 percent processing overhead for each virtual server (10 percent per core). The final server configuration performance will depend on how the virtual servers are deployed on the host platforms (for planning purposes, there should be no more virtual server core than available physical host platform core—do not over allocate).

Looking at figure E.3, the baseline seconds for each tier are from the figure E.2 workflow loads analysis. Adjusted seconds column accommodates the slower E5-2667 processor speeds. Required core calculations include virtual server 20 percent processing overhead (1.2 × 60 sec/adjusted sec). The completed analysis in figure E.3 shows the peak virtual server machine requirements (WTS = 3VM [2+1], WebIn = 2 VM[1+1], WebPub = 2 VM [1+1], DBMS = 2 VM[1+failover]).

The server platform configuration drawing in figure E.3 shows how these virtual server machines should be deployed on the Xeon E5-2667 host platforms for optimum availability. The primary host machine can satisfy peak throughput requirements if the backup machine were to fail. If the primary machine fails, the backup machine can be configured to reboot and recover the primary machine configuration. There are extra host platform core available that can be used for development, testing, and staging virtual server machines as required.

Total hardware server cost includes two host server platforms, with each processor chip configured to host three virtual server machines. Cost per server is $17,185. Cost for deploying virtual servers on four 6-core processor chips is an additional $8,000 (4 × $2,000 per chip). Total hardware cost (including the additional virtual server machines) is $42,370.

The city of Rome year 2 network suitability analysis

The City of Rome year 2 network suitability analysis was shown in figure 9.26. Network upgrade recommendations are provided in figure 9.27. Year 2 software configuration is shown in figure 9.28, and the virtual server platform loads analysis is completed in figure 9.30. Virtual server environment would require a total of four Xeon E5-2667 12-core (2-chip) 2900 MHz host platforms (fourth host machine was required for high-availability failover). Total cost for server platforms was $84,740.

Platform Selection Platform Configuration	SRint2006 per Core	Total Chips
Xeon E5-2667 12 core (2 chip) 2900 MHz	41.8	2

Server Platform tier	Baseline sec	Adjusted sec	Percent Rollover	Required Core*	Core/ VM	Primary VM	Backup VM	Min RAM per VM	VM CPU Utilization
WTS: WTS Platform Tier	98.8	106.5	80%	2.1	2	2	1	19 GB	35%
WebIn: Internal Mapping Tier	50.9	54.9	80%	1.1	2	1	1	5 GB	27%
WebPub: Public Mapping Tier	30.0	32.3	80%	0.6	2	1	1	5 GB	16%
DBMS: Database Tier	29.8	32.1	80%	0.6	2	1	1	19 GB	32%
* Includes 10% processing overhead per virtual server core					Total core =	10	8		

Platform Selection Platform Configuration	Platform Core/node	Total Virtual core	Physical Nodes	Total Servers	RAM per Node	Cost per Server	Total VM	Total Cost
Xeon E5-2667 12 core (2 chip) 2900 MHz	12	18	1.50	2	128 GB	$17,185	$8,000	$42,370
Storage pricing estimates: Volume of data gigabites + 50% (data indexing) x $ per GB at required RAID level (see data storage estimates)								

Server Platform Configuration

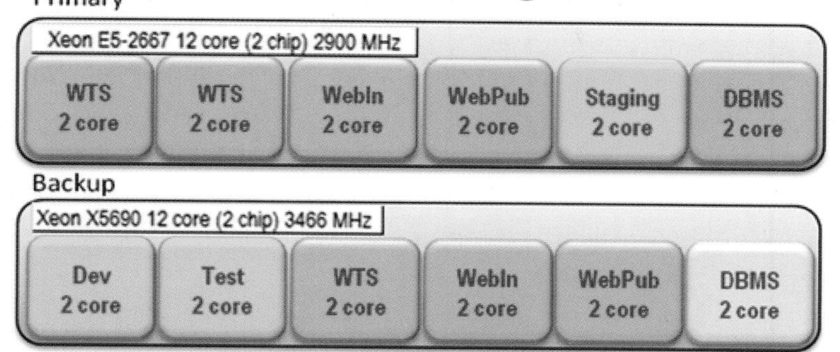

Virtual Server Pricing

Vmware	Price/ chip	MaxCore/ chip
Standard	$2,000	6
Advanced	$4,500	12

$8,000 (4 chip x $2,000)

Figure E.3 **Rome year 1 high-availability virtual server platform selection and pricing.**

Year 2 high-availability virtual server configuration with public web services deployed on Amazon cloud

The second design solution hosts the public web services in the Amazon cloud, reducing the number of required host platforms within the Rome data center. Figure E.4 shows the city network workflow loads without the public web services. WTS and WebIn platform server loads will remain the same. Removing the WebPublic server loads (hosted on Amazon EC2) will reduce the DBMS loads by over 11 percent (102.8/91.2).

The Rome data center peak Internet traffic requirements would also be reduced by 47.2 Mbps, which could result in 50 percent reduction in data center Internet gateway network bandwidth requirements.

City of Rome Year 2	User Environment					Bandwidth Mbps	NW %Cap	Platform Service Times Arc12 4 core (1 chip) Baseline				
Types of Workflows	Peak Concurrent		DPM/TPM		Network		Traffic					
Standard	Users	TPH	per Client	Total	Mbps		Mbpd	Client	WTS	WebIn	WebPub	DBMS
LAN_Local Clients		Services	87 Clients	LAN = 62.4 Mbps		1000	6%	Platform Service Times (seconds)				
GISDeskEditor	19		10.00	190	17.417		5.500	0.190				0.021
GISDeskView	49		10.00	490	24.500		3.000	0.096				0.011
GISDeskBA	18		10.00	180	16.500		5.500	0.190				0.021
WebInternal		8,400	6.00	140	3.967		1.700	0.036		0.104		0.014
BatchAdmin	1		249.29	249						0.157		0.029
WAN_Clients			136 Clients	WAN = 18.8 Mbps		45	42%					
Site 2_Operations			66 Clients	Trafic = 9.7 Mbps		24	40%					
RemoteGISView	66		10.00	660	7.810		0.710	0.036	0.190			0.021
WebInternal		4,000	6.00	67	1.889		1.700	0.036		0.104		0.014
Site 3_Freeberg			30 Clients	Trafic = 3.9 Mbps		12	32%					
RemoteGISView	30		10.00	300	3.550		0.710	0.036	0.190			0.021
WebInternal		700	6.00	12	0.331		1.700	0.036		0.104		0.014
Site 4_Willsberg			40 Clients	Trafic = 5.2 Mbps		12	43%					
RemoteGISView	40		10.00	400	4.733		0.710	0.036	0.190			0.021
WebInternal		1,000	6.00	17	0.472		1.700	0.036		0.104		0.014
Internet_Clients			182 Clients	Internet = 23.0 Mbps		90	26%					
Public_Web Services						90						
WebPublic							1.700	0.036			0.060	0.007
Site 5_Perth			2 Clients	Trafic = 0.3 Mbps		1.5	19%					
RemoteGISView	2		10.00	20	0.237		0.710	0.036	0.190			0.021
WebInternal		100	6.00	2	0.047		1.700	0.036		0.104		0.014
Site 6_Wawash			40 Clients	Trafic = 5.0 Mbps		12	41%					
RemoteGISView	40		10.00	400	4.733		0.710	0.036	0.190			0.021
WebInternal		500	6.00	8	0.236		1.700	0.036		0.104		0.014
Site 7_Jackson			20 Clients	Trafic = 2.6 Mbps		6	43%					
RemoteGISView	20		10.00	200	2.367		0.710	0.036	0.190			0.021
WebInternal		400	6.00	7	0.189		1.700	0.036		0.104		0.014
Site 8_Petersville			60 Clients	Trafic = 7.6 Mbps		18	42%					
RemoteGISView	60		10.00	600	7.100		0.710	0.036	0.190			0.021
WebInternal		1,000	6.00	17	0.472		1.700	0.036		0.104		0.014
Site 9_Rogerton			60 Clients	Trafic = 7.6 Mbps		18	42%					
RemoteGISView	60		10.00	600	7.100		0.710	0.036	0.190			0.021
WebInternal		1,000	6.00	17	0.472		1.700	0.036		0.104		0.014
Total Throughput	405	17,100										

Total baseline platform loads								
Client	WTS		WebIn		WebPub		DBMS	
Users	Users	Sec	Users	Sec	Users	Sec	Users	Sec
19.0							19.0	4.0
49.0							49.0	5.4
18.0							18.0	3.8
23.3			23.3	14.6			23.3	2.0
			1.0	39.1			1.0	7.2
66.0	66.0	125.4					66.0	13.9
11.1			11.1	6.9			11.1	0.9
30.0	30.0	57.0					30.0	6.3
1.9			1.9	1.2			1.9	0.2
40.0	40.0	76.0					40.0	8.4
2.8			2.8	1.7			0.0	0.2
2.0	2.0	3.8					2.0	0.4
0.3			0.3	0.2			0.3	0.0
40.0	40.0	76.0					40.0	8.4
1.4			1.4	0.9			1.4	0.1
20.0	20.0	38.0					20.0	4.2
1.1			1.1	0.7			1.1	0.1
60.0	60.0	114.0					60.0	12.6
2.8			2.8	1.7			0.0	0.2
60.0	60.0	114.0					60.0	12.6
2.8			2.8	1.7			0.0	0.2
451.5	318.0	604.2	48.5	68.8			444.2	91.2

Figure E.4 **Rome year 2: Virtual server platform loads analysis without WebPublic services.**

The selected virtual server host platforms remain the same. The virtual server platform sizing analysis and pricing is provided in figure E.5.

The WTS, WebIn, and DBMS platform tier requirements will be the same as identified in figure 9.30, with the DBMS utilization reduced to 57 percent. The virtual server environment would require a total of three Xeon E5-2667 12-core (2-chip) 2900 MHz host platforms. Total cost for server platforms is reduced to \$63,555 (3 × \$17,185 + 6 × \$2,000).

The City of Rome WebPublic services will be deployed on the Amazon EC2 cloud. City of Rome worked with the Amazon cloud provider to identify pricing information for use in their design analysis. Pricing can vary over time, so it is important to work with the cloud vendor to understand what this cost might be before making a final design decision.

Amazon EC2 pricing is published by Amazon web services on the web. For the purposes of this exercise, we provide a sample of Amazon reserve instance pricing in figure E.6. There are several Amazon Machine Instance (AMI) configurations to choose from, depending on performance, capacity, and scalability requirements.

Platform Selection Platform Configuration	SRint2006 per Core	Total Chips
Xeon E5-2667 12 core (2 chip) 2900 MHz	41.8	2

Server Platform tier	Baseline sec	Adjusted sec	Percent Rollover	Required Core*	Core/ VM	Primary VM	Backup VM	Min RAM per VM	VM CPU Utilization
WTS: WTS Platform Tier	604.2	651.2	80%	13.0	2	9	1	19 GB	65%
WebIn: Internal Mapping Tier	68.8	74.1	80%	1.5	2	1	1	5 GB	37%
DBMS: Database Tier	91.2	98.3	80%	2.3	4	1	1	38 GB	57%
* Includes 10% processing overhead per virtual server core					Total core =	24	8		

Platform Selection Platform Configuration	Platform Core/node	Total Virtual core	Physical Nodes	Total Servers	RAM per Node	Cost per Server	Total VM	Total Cost
Xeon E5-2667 12 core (2 chip) 2900 MHz	12	32	2.67	3	128 GB	\$17,185	\$12,000	\$63,555
Storage pricing estimates: Volume of data gigabtes + 50% (data indexing) x \$ per GB at required RAID level (see data storage estimates)								

Server Platform Configuration

Primary

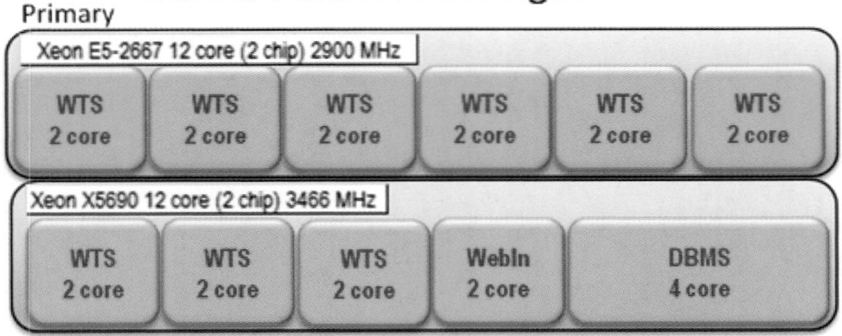

Failover

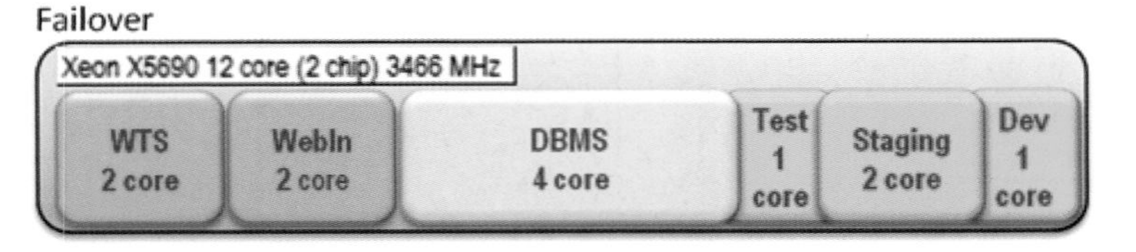

Figure E.5 **Rome year 2 high-availability virtual server platform selection and pricing without WebPublic services.**

We will choose to use the AMI HM Extra Large Instance 2-core (6.5 compute units [CU]) 17.1 GB virtual servers for hosting our public web services. This platform provides the best performing server configuration for our GIS web publishing environment. Server throughput performance benchmarks are identified by Amazon CU. Our server selection provides the fastest available 2-core server solution, which will give optimum web services throughput capacity at minimum software license cost.

Pricing for each virtual server AMI includes a fixed base price ($1,283) plus a variable rate ($0.18 per hour). The base price reserves the purchased AMI for the three year term, and the variable rate is paid for AMI use during the reserve period. You are charged the variable rate only when the AMI is operational (you can choose to shut down an AMI for periods of the day when not required to satisfy operational needs).

The following additional sample charges would be included in the Amazon services monthly billing:

- Internet data transfer IN—$0.00 per GB
- Data transfer OUT—$0.12 per GB/month (first GB per month free)
- Data transfer between Amazon regional data centers—$0.01 per GB
- Public and Elastic IP and Elastic Load Balancing data transfer—$0.01 per GB in/out
- Elastic load balancing—$0.025 per hour
- Amazon Web Services (AWS) import/export data-loading service - $80 per storage device + $2.49 per data-loading-hour
- S3 storage—$0.11 per month per GB

This may not be a complete list of Amazon operational charges. It is important to review and understand your cloud vendor pricing policies and evaluate these costs for accurate pricing.

The Amazon EC2 website provides relative performance information (CUs) on their available instance types that can be used to define AMI performance. Our selected 2-core virtual machine throughput is 6.5 CUs. This CU value can be translated to a SPECrate_int2006 baseline throughput of roughly 58.6, which gives a per-core baseline value of 29.3 that we can use in our platform sizing analysis.

The Amazon cloud virtual server platform sizing analysis and AMI pricing is provided in figure E.7. The Public Mapping Tier baseline processing load (100.0 seconds) is obtained from figure 9.30. The processing load is adjusted based on the selected AMI to 153.8 seconds (100.0 × 45/29.3). The data will be deployed to a file geodatabase supported on Amazon Elastic Load Balancing (ELB) storage for each EC2 AMI server (no additional processing loads). The Amazon web mapping cloud configuration will require two AMI virtual server machines during peak processing loads (peak utilization = 77 percent).

Amazon Reserved Instances (3 year term)	Base	rate
High Memory Reserved Instances		
AMI HM Extra Large Instance 2 core (6.5 CU) 17.1 GB	$1,283	$0.18
AMI HM Double Extra Large Instance 4 core (13 CU) 34.2 GB	$2,566	$0.35
AMI HM Quadrupal Extra Large Instance 8 core (26 CU) 68.4 GB	$5,132	$0.70
Standard Reserve Instances		
AMI Extra Large Instance 4 core (8 CU) 15 GB	$2,000	$0.31
AMI Small Instance 1 core (1 CU) 1.7 GB	$250	$0.04
AMI Large Instance 2 core (2 CU) 7.5 GB	$1,000	$0.16

Figure E.6 **Sample Amazon Cloud AMI pricing.**

Pricing for the two Amazon AMIs is roughly $12,027. This includes a base price ($1,283) plus the variable rate ($0.18 per hour) for a total of 26,280 hours (24 hours per day × 365 days per year × 3 years) for each server. Additional costs can be computed based on data size and estimated transfer rates.

- Data in (100 GB data source, 10 GB monthly updates)
  - $85 initial data load (AWS Import/Export service for 100 GB data)
  - $396 S3 storage (100 GB × $0.11/month × 36 months)
- Data out (100,000 peak transactions per hour, average 300,000 transactions per day)
  - 1.2 Mbpd traffic per transaction (36 GB per day, 1,080 GB per month)
  - $130 per month (1,080 GB @ $0.12)
  - $4,666 over 3 years ($130 × 36 months)
- Elastic Load Balancing (ELB) costs
  - $ 648 total hour cost (24 hrs/day for 3 years @ $.025/hr)
  - $ 388 traffic cost (1,350 GB/month @ $.008/GB for 3 years)

Total cost estimate over the three year period would be $18,210. Pricing includes cost of two Amazon AMIs ($12,027), data upload and S3 storage costs ($481), data out costs ($4,666), and elastic load balancing ($1,036).

Police department network bandwidth suitability analysis

Figure E.8 shows the police department user requirements and network suitability assessment. Police LAN includes the GIS desktop editors and viewers which can use the custom workflow performance targets identified in figure 9.19. An administrative batch process workflow

Platform Selection Platform Configuration	SRint2006 per Core	Total Chips
AMI HM Extra Large Instance 2 core (6.5 CU) 17.1 GB	29.3	1

Server Platform tier	Baseline sec	Adjusted sec	Percent Rollover	Required Core*	Core/ VM	Primary VM	Backup VM	Min RAM per VM	VM CPU Utilization
WebPub: Public Mapping Tier	100.0	153.8	80%	3.1	2	2	0	5 GB	77%
* Includes 10% processing overhead per virtual server core					Total core =	4	0		

Platform Selection Platform Configuration	Platform Core/node	Total Virtual core	Total Servers	Cost per Server	Service rate	Total hours	Total per VM	Total Cost
AMI HM Extra Large Instance 2 core (6.5 CU) 17.1 GB	2	4	2	$1,283	$0.18	26,280	$6,013	$12,027

Figure E.7 **Rome year 2 Amazon cloud virtual server platform selection and AMI server pricing.**

is also included for database maintenance during peak operations. The police department is connected to an Internet wireless service provider through a T-1 (1.5 Mbps) connection. Police vehicles communicate over 56 Kbps wireless connections to towers located throughout the city. Mobile communications traffic is well within the available bandwidth.

Police server workflow platform loads analysis

The next step is to update the workflow software installation. The police department has decided to use a single server platform. Figure E.9 shows the police workflow software configuration. The software is configured on three virtual server machines which include Batch (single-core BatchAdmin batch processing service), Mobile (single-core mobile synchronization service), and DBMS (two-core database server).

The mobile GIS client is a stand-alone, loosely connected device that operates from a local data cache. Central police data updates are provided through the wireless mobile synchronization service. The baseline mobile synchronization service handles point, polygon, and simple attribute changes with the central police web server. Baseline service times are very light for each data synchronization (0.034 sec).

The BatchAdmin service provides resources for a single IT batch processing task running during peak utilization periods. The DBMS provides a common geodatabase data source for the police business layers.

Police Department	User Environment					Bandwidth	NW
Types of Workflows	Peak Concurrent		DPM/TPM		Network	Mbps	%Cap
Standard	Users	Client	per Client	Total	Mbps		Traffic Mbpd
LAN_Local Clients		Services	24 Clients	LAN = 25.7 Mbps		100	26%
GISDeskEditor	15		10.00	150	13.750		5.500
GISDeskView	9		10.00	90	4.500		3.000
BatchAdmin		1	222.07	222	7.402		2.000
WAN_Clients			20 Clients	WAN = 0.1 Mbps		1.5	5%
MobileClient	20		6.00	120			
MobileService		20	3.00	60	0.076		0.076

Figure E.8 **Police network suitability analysis.**

Police Department			Software Configuration						Platform Service Times			
			User Environment						Arc12 4 core (1 chip) Baseline			
Types of Workflows	Peak Concurrent		Desktop	Web	Server	Data Source						
Standard	Users	Client	Client	Web	SOC	SDE	DBMS	Data Source	Client	Batch	Mobile	DBMS
LAN_Local Clients		Services	Client	Batch	Batch	Default	DBMS		Platform Service Times (seconds)			
GISDeskEditor	15		Default			Default	Default	SDE_DBMS	0.190			0.021
GISDeskView	9		Default			Default	Default	SDE_DBMS	0.096			0.011
BatchAdmin		1		Batch	Batch	Default	Default	SDE_DBMS		0.157		0.029
WAN_Clients												
MobileClient	20		Default					SDE_DBMS	0.166			
MobileService		20	Default	Mobile	Mobile	Default	Default	SDE_DBMS	0.036		0.033	0.001

Figure E.9 **Police workflow software configuration.**

Figure E.10 shows the workflow loads, platform selection, and cost analysis for the police department year 2 deployment. Peak baseline processing time for the police batch server is 46.6 seconds, Mobile server is 2.0 sec, and the DBMS is 12.8 sec. We selected the Xeon E5-2647 4-core server as the police host platform.

The police department virtual server platform sizing analysis and pricing is provided in figure E.11.

The Mobile synchronization service 1-core virtual servers peak at 2 percent utilization, the two batch 1-core virtual servers peak at about 42 percent (they could easily handle two concurrent batch processes), and the database 2-core virtual server peaks at 12 percent utilization. Virtual server environment with two Xeon E5-2637 4-core (2-chip) 3000 MHz host platforms provide fully redundant high-availability operations. Total cost for server environment is $36,770 (2 × $14,385 + 4 × $2,000).

Police Department	User Environment					Bandwidth Mbps	NW %Cap	Platform Service Times Arc12 4 core (1 chip) Baseline				Total baseline platform loads						
Types of Workflows	Peak Concurrent		DPM/TPM		Network		Traffic					Client	Batch		Mobile		DBMS	
Standard	Users	Client	per Client	Total	Mbps		Mbpd	Client	Batch	Mobile	DBMS	Users	Users	Sec	Users	Sec	Users	Sec
LAN_Local Clients		Services	24 Clients	LAN = 18.3 Mbps		100	18%	Platform Service Times (seconds)										
GISDeskEditor	15		10.00	150	13.750		5.500	0.190			0.021	15.0					15.0	3.2
GISDeskView	9		10.00	90	4.500		3.000	0.096			0.011	9.0					9.0	1.0
BatchAdmin		1	297.01	297					0.157		0.029		1.0	46.6			1.0	8.6
WAN_Clients			20 Clients	WAN = 0.1 Mbps		1.5	5%											
MobileClient	20		6.00	120				0.166				20.0						
MobileService		20	3.00	60	0.076		0.076	0.036		0.033	0.001	20.0			20.0	2.0	20.0	0.1
Total Throughput	44	21										64.0	1.0	46.6	20.0	2.0	45.0	12.8

Figure E.10 **Police year 2: Virtual server platform loads analysis.**

Platform Selection Platform Configuration	SRint2006 per Core	Total Chips
Xeon E5-2637 4 core (2 chip) 3000 MHz	46.3	2

Server Platform tier	Baseline sec	Adjusted sec	Percent Rollover	Required Core*	Core/ VM	Primary VM	Backup VM	Min RAM per VM	VM CPU Utilization
Batch: Internal Mapping Tier	46.6	45.4	80%	0.8	1	2	0	2 GB	42%
Mobile: Public Mapping Tier	2.0	1.9	80%	0.0	1	1	1	2 GB	2%
DBMS: Database Tier	12.8	12.5	80%	0.2	2	1	1	21 GB	12%
* Includes 10% processing overhead per virtual server core					Total core =	5	3		

Platform Selection Platform Configuration	Platform Core/node	Total Virtual core	Physical Nodes	Total Servers	RAM per Node	Cost per Server	Total VM	Total Cost
Xeon E5-2637 4 core (2 chip) 3000 MHz	4	8	2.00	2	48 GB	$14,385	$8,000	$36,770
Storage pricing estimates: Volume of data gigabites + 50% (data indexing) x $ per GB at required RAID level (see data storage esimates)								

Platform Configuration

Primary server

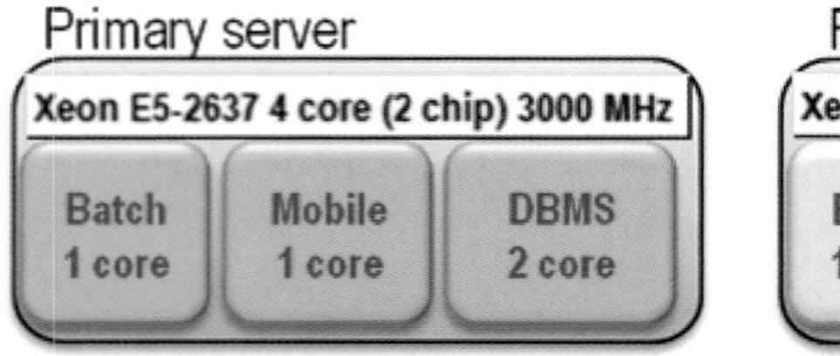

Failover server

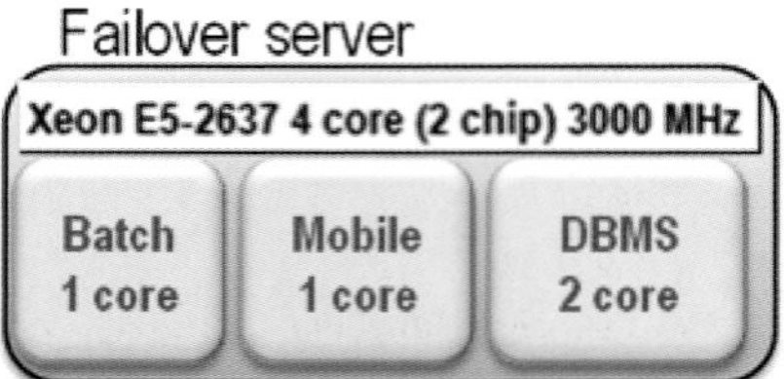

Figure E.11 **Police year 2 high-availability virtual server platform selection and pricing.**

Lexicon

This lexicon contains definitions of the most important functions that your GIS software should provide. Many of the GIS function entries include examples of their use. This is not a comprehensive list of GIS functions, but rather an accessible, convenient subset of the most commonly used functions, which will help you prepare information product descriptions (IPDs) and functional specifications.

You will discuss GIS functions during several stages in the GIS planning process:

- During the technology seminar (chapter 5), to ensure that the planning team shares a common vision of what a GIS is and to encourage the use of a common vocabulary to assist communication
- During the development of IPDs (chapter 6), to identify the GIS functions necessary to create an information product
- During conceptual system design (chapters 7–9), to assess function use to help evaluate system requirements in preparation for making procurement recommendations (and planning implementation)

The lexicon is designed to boost your planning efforts in at least three ways:

1. The lexicon will broaden your knowledge of the full range of GIS functions that should be available in a comprehensive system. (Most GIS users access only about 10 percent of the functions available to them on a daily basis.)
2. It will enable you to write a description of your needs in a way that will be understood by any user or vendor. Users and vendors of GIS software tend to become well-versed in the terminology of their particular software. Different software, however, may have different terms for the same function or concept. Therefore, a software-independent description—not tied to any brand name—should facilitate communication even when terminology varies.
3. It will allow non-GIS users to appreciate the full capabilities of a GIS. Many of those involved in GIS planning, as well as senior managers, clients, and members of an organization's information technology department, may be new to GIS software. The lexicon will help you help them understand its potential to serve their interests.

More GIS terminology can be found on the Esri GIS Dictionary site, http://support.esri.com/en/knowledgebase/gisdictionary/browse

The name of each function is followed by a brief description of the process or ability you should expect from it, via the software you choose. Functions that are considered highly complex are flagged by an asterisk (*).

The lexicon is divided into the following sections or task categories; you'll find each function under the overall task it most often supports:

Data input
Data storage, data maintenance, and data output
Query
Generating features
Manipulating features
Address locations
Measurement
Calculation
Spatial analysis
Surface interpolation
Temporal analysis
Visibility analysis
Modeling
Network analysis

Data input

Digitizing

The process of converting point and line data from source documents into a machine-readable format. Manual methods employing a digitizing table or tablet are widely used, but, increasingly, these methods are being replaced by scanning, automatic line following methods, and transferring data files already in digital format. Many users, however, still use a manual digitizing technique for small amounts of data or when other methods are too expensive. Digitized data often needs editing or reformatting.

Scanning

The process of creating an electronic photocopy of a paper map or document. You can use a flatbed or drum scanner, depending on the size of your map or document and the resolution you require. Scanning results in data that is in raster format (see "Raster-to-vector and vector-to-raster conversion," later in this section). The data layer produced will contain all of the detail on the input map or image—including features you may not want to collect. For this reason, post-processing of scanned data is common. Scanned raster images are useful as a background for vector data, as they provide good spatial context and assist interpretation.

Keyboard input

The process of manually typing alphanumeric data into machine-readable form; used infrequently because the dangers of human error are high, and inputting multiple entries and checking are labor-intensive. Occasionally, keyboard entry is used to label data taken from hardcopy maps or, less frequently, to enter coordinates for map corrections or very simple maps.

File input

The process of entering tabular data or text from ASCII or word-processor-generated files that have already been manually typed and checked.

File transfer

Allows you to input data that has been created with some other software or system. The data file being transferred may come from external, commercial data providers or from other systems within your own organization (for example, GPS or CAD). Data can be transferred from disk, CD-ROM, the Internet, or downloaded directly from field-data collection devices. This data might need to be reformatted to make it compatible with your GIS.

Raster-to-vector and vector-to-raster conversion

Processes that change the format of data to allow additional analysis and manipulation. During vector-to-raster conversion, you should be able to convert both the graphical and topological characteristics of the data. You should also be able to select cell size, grid position, and grid orientation. During raster-to-vector conversion, creation of topology is necessary. Editing and smoothing of data may be necessary to create an effective vector representation.

Data editing and data display

These functions should be possible at any time during digitizing and may apply to points, lines, and labels. On-screen displays or paper plots of errors assist editing. A range of functions for displaying different portions or different features of a dataset is essential. Functions that should be available for edit and display include the following:

- Selecting one or more datasets for display and editing
- Selecting a specific area within a dataset for editing
- Selecting features (points, lines, labels, etc.) to be displayed
- Selecting a feature or set of features to be edited
- Querying the attributes of selected features
- Indicating line ends (nodes) on request
- Rotating, scaling, and shifting one dataset with respect to another
- Deleting all user-selected features, codes, or data types in specified areas
- Deleting a user-selected feature using attributes or the mouse
- Adding a feature using the cursor, keyboard, or mouse
- Moving all or part of a feature
- Creating and modifying areas interactively
- Changing label text or location

Create topology*

The ability to create digitized lines that are intelligently connected to form polygons or networks in two or three dimensions. Ideally, the process is automatic and includes error correction procedures (e.g., "join line endpoints within specified tolerance distances" or "remove small overshoots"). Alternatively, errors could be highlighted on-screen for you to edit and correct using standard graphical editing functions. The ability to perform topological operations on a selected part of the database is desirable, rather than involving the whole database each time.

Edgematching*

An editing procedure used to join lines and areas across map sheet boundaries to produce a single, seamless digital database. The join created by edgematching needs to be topological as well as graphical. An area joined by edgematching should become a single area in the final database, and a line joined by edgematching should become a single line. Edgematching functions should be able to handle small gaps in data, slight discrepancies, overshoots, and missed and double lines. If these gaps and errors cannot be handled automatically, you should be alerted to their presence so that other methods of correction can be applied. Edgematching functions should allow you to set a tolerance limit for automatic editing.

Adding attributes

The process of adding descriptive alphanumeric data to a digital map or to an existing attribute table. Attributes are characteristics of geographic features (points, lines, and areas). They are typically stored in tabular format and linked to the features. When there are a significant number of attributes associated with features (or sometimes for database design reasons), attributes may be stored in separate databases within your GIS or within another database management system (DBMS).

Reformatting digital data

The process of making digitized data or data transferred from another system compatible with your system. Reformatting ensures accessibility or assists conversion to the system software format. Reformatted data should be topologically and graphically compatible with other data in your GIS. You may need to reformat data so that it complies with organizational standards. Reformatting may involve additional digitizing, adding extra labels, automatic editing, and the editing of attribute data.

Automated reformatting is increasingly important as users wish to integrate multiple data formats (from open or proprietary sources) into their GIS directly or make their own data available to others over the web. Interoperability programs are available that allow direct reading and export from multiple sources.

Schema change

The process of automatically combining two classification systems into one acceptable classification according to a set of rules. Usually carried out during the process of automated reformatting. Also called reclassification.

Data storage, data maintenance, and data output

Create and manage database

The process of organizing datasets using good cartographic data structures and data compression techniques. Database creation and management permits easy access to data when a GIS contains several datasets, particularly if some of these are large.

In most cases, data will come from map sheets, images, or documents that cover large, contiguous areas of land. The data from these sheets must be edgematched into a combined database with a consistent data structure. This will allow analysis and query functions to be applied to part or the whole database.

Cache basemaps

Map caches are a place to store reusable data. Caches can be made of multilayer, geoprocessed datasets. Caching enables data to be retrieved very quickly and reduces subsequent platform loads.

Partial cache: Selected part of the full data set is cached

On-demand caching: Geographic features are dynamically rendered and map cache tiles are created during the first query request; future requests use the cached map tiles.

Mixed-mode cache: In image caches, JPEG and transparent PNG32 modes can be used together to enable transparent edges for partial data updates (PNG32) and minimize space requirements (JPEG).

Compact cache: Uses bundles of "tiles" to reduce storage volume and makes it easier to manage and transport caches.

Export and import cache: Allow multiple users and organizations to import to a local workspace (disconnected caching) and collaborate on building a cache

Local caching: Selected feature classes can be cached locally on the desktop for project use

Disconnected caching: Allows desktop users to take caches—no longer tied to the server—into the field.

Edit and display (on output)

The ability to edit and display output map products on-screen is necessary to create effective information products. This process requires a wide variety of functions for editing, layout, symbolization, and plotting. Editing on output includes all of the editing capabilities used on input.

Appropriate symbolization assists presentation of results and simplifies interpretation of data. To facilitate symbolization, your GIS should include a wide variety of symbols that can be used to display points, lines, and areas; the ability to locate and display text and other alphanumeric labels; and the ability to create your own symbols

Symbolize

The process of selecting and using a variety of symbols to represent the features in your database on-screen and on printed output. To create high-quality output from a GIS, you should use a wide variety of symbols to represent the features stored in the database on-screen and on printed output. Functions for symbolizing should permit the following:

- Use of standard cartographic symbols
- Filling areas with patterns of symbols or cross-hatching of different densities
- The representation of point features at different sizes and orientations
- The use of multiple discipline-oriented sets of symbols (for example, geological, electrical, oil, gas, water, and weather).

Plot

The process of creating hard-copy output from your GIS. Printing and plotting functions in your GIS should allow the production of "quick-look" plots on-screen and on paper, spooling or stacking of large print jobs, plotting onto paper or Mylar sheets in a range of sizes, and registration facilities to permit overprinting on an existing printed sheet.

Update

The process of adding new points, lines, and areas to an existing database to correct mistakes or add new information. After the initial creation of a digital database, periodic updates may be necessary to reflect changes in the landscape or area of interest. New buildings and roads may be constructed, and old buildings and roads may be demolished. Quarries or forests may change in their extent or ownership, requiring updating of both the spatial extent and attributes of the data. You may need to update data to correct errors in the data.

Many of the functions for editing and display will be useful for updating, as will functions for heads-up digitizing on-screen. The ability to undo work is important, and there should be transaction logs, backup, and access protection for files during updating.

Browse

Browsing is used to identify and define an area or window of interest that can be used for other functions. During browsing, no modifications to the database should be possible, but you should be able to select areas by specifying a window or central point, and pan and zoom. After identifying an area of interest, you should be able to edit, measure, query, reclassify, or overlay data.

Suppress

Suppression is used to remove features from your working environment so that they are omitted from subsequent manipulation and analysis. As opposed to querying, which is normally used to select features you're interested in, suppression is used to omit features you're not interested in. For example, you might have a data source that contains all the roads in your study area, but you want to work only with the major highways. You can suppress all features other than the major highways so they are excluded from subsequent overlay, display, and plotting operations.

Create list (report)

The creation of lists and reports can be part of the process of generating final or interim information product output. Also, information products themselves may be in the form of tables, lists, and reports. Functions in your GIS should allow you to do the following:

- Create user-specified lists of the results of any function generating alphanumeric output
- Produce subtotals, summary totals, and totals from lists of numbers
- Perform arithmetic and algebraic calculations based on given formulas
- Perform simple statistical operations, such as percentages, means, and modes
- Create list titles and headings in a range of standard formats and easily prescribed custom formats
- Sort data
- Create reports of system errors to permit corrections to be made easily

Create pop-ups

Creating a tailored list of attributes that are associated with features; pop-ups appear when you click a feature in a map.

Serve on Internet

Serving GIS or map data over the Internet usually involves displaying interactive maps that allow users to browse geographic data. In addition, many map servers allow users to view attributes, query the database, and create customized maps on demand.

Query

Spatial query

The process of selecting a subset of a study area based on spatial characteristics. The subset can be used for reporting, further study, or analysis.

Spatial queries are usually implemented by selecting a specific feature or by drawing a graphic shape around a set of features. For example, an irregular study area boundary may be plotted on-screen and all features within this boundary selected, or an administrative area might be selected with a single mouse click and used as a subset of the area for further study.

Querying a database can become complex and involve questions of both spatial and attribute data. For example, "Which properties are on the east side of town?" might be followed by "Which properties have four bedrooms and are available for sale?"

Queries are one of the most commonly used GIS functions. A good system will offer several alternative methods for querying to meet the needs of a range of users.

Attribute query

The process of identifying a subset of features for further study based on questions about their attributes.

Attribute queries are usually implemented using a dialog box that helps build the question or by using a special query language, such as Structured Query Language (SQL). The questions "Which roads have two lanes?"

and "Which properties are zoned residential?" would both result in the selection of a subset of features for further study.

Generating features

Generate features

The ability to create new features and add them to the database. Generating functions should allow features to be defined easily, with no limit on the number of new features that can be added to the database or on the number of points in any position. Names or codes can be attached to new features.

The types of features that you should be able to generate with your GIS include points, lines, polygons, circles, grid cell nets, and latitude-longitude nets.

Generate buffer

The ability to generate zones of specified width around point, line, or area features. Around point and area features, these zones are generally called *buffers,* while zones of interest around line features may be called buffers or *corridors.*

A user-specified buffer distance is used to generate these buffers and corridors, and the system should automatically resolve overlaps and inclusions in cases where features are complex or highly convoluted. Buffers may be necessary both outside and inside area features such as lakes. For point, line, and area features, buffers at different distances (multiple buffers) should be possible. Constant- and variable-width buffers should also be possible, including buffers that intersect each other. Buffer width should be able to be set from the attributes of the features concerned, without operator intervention.

Generate viewshed*

Involves manipulating a digital elevation model to identify areas of the terrain that are visible from one or more viewpoints. The viewpoints may be any point along a line (such as a road) or in a user-defined polygon. Viewshed maps help find well-exposed places for communication towers or more amenable locations for parking lots, for example. Shadow analysis of 3D buildings can now be accomplished.

Generate perspective view*

The ability to generate a three-dimensional block diagram showing the nature of the surface relative to three axes from a digital elevation model. Hidden line removal, hill relief shading, and the ability to plot symbols and cross-hatched areas on the surface plane are desirable to achieve a good quality output.

With scene generation, an advanced form of generating a perspective view, you can generate three-dimensional objects (for example, buildings and trees) and add them to the view. The realistic visualization this provides allows you to dynamically view the scene (perform fly-bys over and under the scene) and dynamically label the features on the passing scene. This can now be carried out from a two-dimensional base using a set of rules—procedural scene creation to produce a three-dimensional view. Quick sketch is available.

Generate elevation cross section*

The ability to generate a graph showing a cross section through a digital elevation model, along a user-defined line of any length or orientation. It is useful if the locations of features that cross the line of section (for example, roads) can be annotated.

Generate graph

The ability to create a graph of attribute data. Graphs are used to display two attributes: one measured along the x-axis and the other along the y-axis. You should be able to illustrate data distribution with symbols, bars, lines, or fitted trend lines. Graphs can be drawn in place of maps or used to supplement them.

Manipulating features

Originally, only two-dimensional features could be manipulated. It is now possible to move 3D representations and use geoprocessing data manipulation capabilities as appropriate.

Classify attributes

Classification is the process of grouping features with similar values or attributes into classes. Many datasets contain a wide range of values. Classifying the data into a number of groups, or classes, for presentation or analysis helps illustration and interpretation. For example, population totals for one-kilometer grid cells may range from zero to several hundred in a study area. Displaying all possible values on a single map, using a different color for each one, could result in a map so multicolored as to render it impossible for a user to interpret. For presentation purposes, up to eight classes are normally used, but for analysis more classes may be appropriate.

Dissolve and merge

Allows the removal of boundaries between two adjacent areas that have the same attributes. A common attribute is assigned to the new larger area. Using these functions, the boundaries between adjacent areas with the same attributes are dissolved to form larger areas; tables containing attribute values from the joined areas are then merged to give one value for the resulting larger area.

This function may be necessary after an edgematching or reclassification operation (although in some cases it may be important to retain the boundaries between areas—for example, administrative or political boundaries).

Line thin*

The ability to reduce data file sizes where appropriate, by reducing line detail after entry. This function reduces the number of points used to define a line or set of lines, in accordance with user-defined tolerances. Some of the points along the line are weeded out to reduce the total number used to represent features. It is important that the general trend and information content of lines be preserved during the process.

Line smooth*

As opposed to line thinning, line smoothing involves adding detail to lines to represent a feature more effectively. Line-smoothing functions use tolerances to smooth lines by adding extra points and reducing the length of individual line segments. The result is a smoother appearance. A number of different functions for line smoothing may be available in a GIS.

Generalize*

A process to reduce the amount of detail when displaying features. Generalization techniques are used to permit effective scale changes and to aid the integration of data from different source scales. A large-scale map (1:50,000) redisplayed on-screen at a smaller scale (1:250,000) would appear cluttered and difficult to interpret without the aid of generalization techniques.

Clip

Allows you to extract features in the database from a defined area. This function is also commonly referred to as *cookie-cutting*. Whether you define the area on-screen with the mouse or by using another feature in the database (such as an administrative area), the result is a new data layer containing only the features of interest within your study area. The original data layer remains unchanged.

Scale change

Involves changing the size at which data is displayed. A scale change is usually performed in the computer rather than at the plotting table. Zoom-in and zoom-out functions should be available, as well as the ability to specify the exact scale at which you want to redisplay data. Line thinning and weeding operations or line smoothing may be incorporated in scale reduction. Line dissolving and attribute merging functions usually need to be invoked prior to broad-range scale reduction. Particular attention should be paid to the legibility of the final product, including labels.

Changing the scale of a dataset before integrating it with other data should be undertaken with caution, as data is best manipulated and analyzed at the scale of collection. As a general rule, if the data is to be used in analysis, you should avoid changing a dataset's scale to more than 2.5 times larger or smaller than the scale of the original source.

Projection change

Allows you to alter the map projection being used to display a dataset. You may need to change the projection of a dataset to enable integration with data from another source. For example, you would do so with data digitized from a map that uses the universal transverse Mercator projection if you wanted it to be overlaid by a data layer using an equal-area cylindrical projection. Your software should provide functions for changing data between a range of common projections or map datums.

Transformation*

The process of converting coordinates from one coordinate system to another through translation (shift), rotation, and scaling. Transformation involves the systematic mathematical manipulation of data: the function is applied uniformly to all coordinates—scaling, rotating, and shifting all features in the output. It is often used to convert data in digitizer units (most often inches) into the real-world units represented on the original map manuscript. Data from CAD drawing files may require transformation to convert from page units to real-world coordinates and permit integration with other data.

Rubber sheet stretch*

Used to adjust one dataset—in a nonuniform manner—to match up with another dataset. If you have two data layers in your GIS that you need to overlay, or one map on a digitizing table that needs to be registered to another map of the same area already in the system, rubber sheet stretching may be necessary. The function allows maps to be fit together or compared. Using common points or known locations as control points, the rest of the data is "stretched" to fit one data layer on the other. Rubber sheet stretching is frequently used to align maps with image data.

Conflate*

Conflation aligns the lines in one dataset with those in another and then transfers the attributes of one dataset to the other. Conflation allows the contents of two or more datasets to be merged to overcome differences between them. It replaces two or more versions of the dataset with a single version that reflects the weighted average of the input datasets. The alignment operation is commonly achieved by rubber sheet stretching.

One of the most common uses of conflation is in transferring addresses and other geocoded information from street network files (for example, TIGER/Line files from the US Census Bureau) to files with more precise coordinates. Many files contain valuable census data but may be deficient in coordinate accuracy. Because attributes are valuable, conflation procedures were developed to transfer the attribute data to a more desirable set of coordinates.

Subdivide area*

The ability to split an area according to a set of rules. As a simple example, given the corner points of a rectangular area, it should be possible to subdivide the area into ten equal-sized rectangles. The boundary of the area may be irregular, however, and the rules applied may be complex. The rules will allow factors such as maximum lot size and road allowances to be taken into account during subdivision planning.

Sliver polygon removal

A sliver polygon is a small area feature that may be found along the borders of areas following the topological overlay of two or more datasets with common features (for example, lakes). Topological overlay results in small sliver polygons if the two input data layers contain similar boundaries from two different sources. Consider two data layers containing land parcels that will be used in a topological overlay. One data layer may have come from an external source—perhaps provided in digital format by a data supplier. The other data layer may have been digitized within the organization. After they are overlaid, it is likely that small errors in the location of parcel boundaries will appear as sliver polygons—small thin polygons along the boundaries.

Automatic functions to remove sliver polygons are available and are commonly incorporated within both topological overlay and editing functions. You should have control of the algorithms used for sliver removal. In particular, you should be able to control the assignment algorithm that will determine to which neighboring polygon a sliver is assigned or how it is corrected.

Address locations

Address match

The ability to match addresses that identify the same place but may be recorded in different ways. This function can eliminate redundancy in a single list (for example, a list of retail store customers), but is more frequently employed to match addresses on one list to those on one or more other lists. Address matching is often used as a precursor to address geocoding. The user can specify various levels of matching probability.

Address geocode

Address geocoding is the ability to add point locations defined by street addresses (or other address information) to a map. Address geocoding requires the comparison of each address in one dataset to address ranges in the map dataset. When an address matches the address range of a street segment, an interpolation is performed to locate and assign coordinates to the address. For example, a text-based data file containing customer addresses can be matched to a street dataset. The result would be a point dataset showing where customers live. The resulting points must be topologically integrated with the database and usable as new features in the database. The new features should be usable by other system functions in combination with the rest of the database. Address geocoding can now be carried out worldwide.

Measurement

Measure length

The ability to measure the length of a line. Measurements in a vector database may be calculated automatically and stored as part of the database. In this case, length can be retrieved from the database with simple query operations. In other cases, measurements may be calculated after you click on a source feature of interest. For example, you might select two locations from your on-screen map, then ask that the distance between them be calculated.

Measure perimeter

The ability to measure the perimeter of an area. Measurements in a vector database may be calculated automatically and stored as part of the database. In this case, perimeter can be retrieved from the database with simple query operations. In other cases, measurements may be calculated after you click a source feature of interest. For example, you might select a field from your on-screen map, then ask that the perimeter be calculated.

Measure area

The ability to measure the area of a polygon. Measurements in a vector database may be calculated automatically and stored as part of the database. In this case, area can be retrieved from the database with simple query operations. In other cases, measurements may be calculated after you click a source feature of interest. For example, you might select a land parcel from your on-screen map, then ask that the area of the parcel be calculated.

This function of your software should also have the capability of calculating the area of a user-defined polygon. User-defined polygons may subdivide existing area features in the database. In this case, only that part of the area feature within the user-defined polygon should be measured. The function should measure interior areas contained within polygons (for example, islands within lakes) and subtract them from the overall area of the feature. In other words, it should be possible to implement three levels of "stacking" in area calculations without operator intervention. For example, you should be able to measure the area of land mass in a polygon that includes a lake that, in turn, contains an island, on which there is a pond.

Measure density

The ability to calculate the density of features in a given area. Now operable in 3D.

Measure volume*

The ability to measure the amount of three-dimensional space occupied by a feature. This includes the ability to measure total floor area in multi-story buildings. Volume measurements can be calculated when surface digital elevation models of features have been incorporated into the database (for example, you could measure the volume of a mountain, the volume of a lake, or the volume of an aquifer).

Calculation

Calculate centroid*

This function calculates the centroid of an area (or set of areas or grid cells) within a user-defined region. It generates a new point at the centroid and automatically allocates a sequential number to each centroid in the region. A useful technique for labeling polygons created during digitizing, centroid calculation is often performed automatically on new areas created by dissolve and merge or by overlay operations.

Calculate bearing

The ability to calculate the bearing (with respect to true north) between two or more points in a database. This is a geometric calculation based on the relationships between features. You should be able to perform this calculation independently or in combination with other arithmetic, algebraic, or geometric calculations in macroprograms and iterative procedures.

Calculate vertical distance or height

The ability to calculate the vertical distance (height) between two points in a digital elevation model. The calculation of vertical distance between two points should be possible wherever the points are located in the region covered by the digital elevation model.

Calculate slope

Calculation of slope (change in surface value) is the ability to calculate the slope along lines, or the average slope of an area.

Calculate aspect*

The ability to calculate the compass direction toward which a slope faces. This function requires a digital elevation model and a user-specified area. The average aspect of the region should be calculated, weighted by the amount of land in each aspect category.

Calculate angle and distance*

The function that can generalize the shape of a linear feature into a set of angles and distances from a starting point. The user should be able to set angular increments and constrain the calculation to any known point along the linear feature.

Calculate location from a traverse*

The ability to calculate the route and endpoint of a traverse, given a starting point and directions and distances of travel. It should be possible to enter the resulting route and endpoint (a point or grid cell) into the database.

Arithmetic calculation

The ability to perform operations such as addition, subtraction, multiplication, and division. You should be able to perform arithmetic calculations independently, perform arithmetic calculations in combination with algebraic and geometric functions, incorporate arithmetic calculations into macroprograms, change variables and components of algorithms, and establish iterative procedures.

Algebraic calculation

The ability to perform operations based on logical expressions. You should be able to perform algebraic calculations independently, perform algebraic calculations in combination with arithmetic and geometric functions, incorporate algebraic calculations into macroprograms, change variables and components of algorithms, and establish iterative procedures.

Statistical calculation

Statistical functions perform simple statistical analyses and tests on the database.

Increasingly, statistical functions are common in GIS software programs; however, for more sophisticated analysis, data may have to be transferred to other statistical packages. Statistical functions should allow you to calculate mean, median, standard deviation, variance, percentiles, cross-tabulations, and regression.

Spatial analysis

Graphic overplot

The ability to superimpose one map on another and display the result on-screen or as a plot to see the intersection of the datasets. When you use graphic overplot, the datasets are not integrated in the database and no new datasets are created. This function merely produces a visual impression of the interrelationships between two (or more) datasets.

Graphic overplotting is commonly used to combine thematic data layers to give a context for interpretation. For example, you might display the boundary of your study area, the roads within the areas, land-use polygons, and rivers and lakes before performing queries or other analysis. Graphic overplotting can also be used to display the results of analysis in a way that aids interpretation. A land-use dataset might be overplotted on a landscape surface to provide a three-dimensional visualization of the changes in land use across a study area.

Topological overlay*

Topological overlay of one map on another will produce new data as a result of the combination of two input data layers. The attributes of the two input layers will be combined into a new set of attributes for the accompanying output layer.

Three types of topological overlay are frequently used:

Point in polygon overlay allows you to superimpose a set of points on a set of polygons, determine which polygon (if any) contains each point, and add the results to the database as attributes of the points. If a point is contained within a polygon, the attributes of that polygon are added to the point.

Line on polygon overlay allows you to superimpose a set of lines on a set of polygons. Lines are broken at intersections with polygon boundaries, and the attributes of the polygon that each segment of the line crosses are added to the attributes of that segment.

Polygon on polygon overlay allows you to superimpose two polygon datasets. The result is a topologically integrated version of the two input datasets that can be used to create a new output map or for further analysis. Polygons in the output map will have attributes from both of the input maps.

Adjacency analysis*

The ability to identify areas that are next to (adjacent to) each other, particularly those that share a common boundary.

Connectivity analysis*

The ability to identify areas or points that are (or are not) connected to other areas or points by tracing routes along linear features.

Nearest neighbor search*

The ability to identify individual or sets of points, lines, or areas that are nearest to other points, lines, or areas specified by location or attributes.

Correlation analysis*

The ability to compare maps that show the same area, but that represent conditions in different time periods. Correlation can be a very useful management tool. Quantifying and explaining the differences between two maps requires determining and comparing the differences between them. Correlation is one method for doing this. It may involve using overlay techniques and statistical functions.

Linear referencing*

The ability to associate multiple sets of attributes with any portion of a linear feature. These attributes can be stored, displayed, queried, and analyzed without affecting the underlying linear data's coordinates. Linear referencing models linear features using routes and events.

A route represents a linear feature such as a city street, highway, or river. Routes contain measures that describe distance along them. These measures provide an explicit location for data that describes parts of the route. The attributes associated with any occurrence along the linear features are known as events. Events are stored in a tabular database rather than with the data's geometry; therefore, they do not affect the underlying spatial data. These events are accessed as needed from the tabular database.

Linear referencing allows the computation of locations of events on linear features based on an event table for which distance measures are available.

Temporal analysis

Time-enabled data can be authored with time properties of each layer. You can use this information to change the display of your map or perform temporal analysis. A time slider tool provides controls that allow you to visualize temporal data using a graphic slider.

Surface interpolation

Interpolate spot height*

The ability to predict the height of any point in an area from a digital elevation model. A new point is generated with height as an attribute.

Interpolate spot heights along a line*

The function that can predict heights along lines using a digital elevation model. For example, if you have a digital elevation model and a hydrology network, interpolation can be used to generate points along streams at fixed increments of height (for example, ten feet) above a given point on the stream. The same technique could be used with other networks, such as roads or pipelines.

Interpolate isoline (contour)*

The ability to generate lines showing equal elevation from a set of regularly or irregularly spaced point values. If the values are height values from a digital elevation model, contours will be created. If the point values represent pressure readings, isolines are created.

Interpolate watershed boundaries*

The ability to generate areas of drainage using a digital elevation model and a hydrology network. Many terms are used to refer to the areas of drainage, including drainage basin, watershed, basin, catchment area, and contributing area.

Visibility analysis

Line of sight*

These functions compute the points, parts of lines, and sections of polygons that are visible along a line between a given target and a point of observation. Line-of-sight calculation requires a surface. Commonly, surface data comes from a digital elevation model. If you were physically located at one point in your dataset (say on top of a mountain), a line-of-sight calculation will establish whether you would be able to see from that point to a target point (such as a lookout tower on another mountain peak some distance away).

Generate viewshed*

Generating a viewshed involves manipulating a digital elevation model to identify areas of the terrain that are visible from one or more viewpoints. (Also in "Generating features" category.)

Modeling

Arithmetic modeling*

Used to add, subtract, multiply, or divide the values of one or more input datasets to calculate the values for a resulting dataset.

Weighted modeling*

Allows you to assign weighting factors to individual datasets according to a set of rules, and to overlay those datasets and perform reclassify, dissolve, and merge functions on the resulting concatenated dataset. This may be done to identify regions with specific characteristics (for example, zones suitable for development). In this instance, proximity to market may be given a higher weight in the modeling process than slope or aspect characteristics of the land.

Network analysis

Shortest route*

Functions to determine the shortest or minimum value path between two points or sets of points on a network. The minimum value may be expressed in terms of, for example, cost or time. When complex network analysis is not required, shortest route functions may be sufficient for many users. Shortest route can be used on any type of network data, including transportation, river, pipeline, or cable networks. Routing may now be accomplished worldwide.

Network analysis*

Functions that allow you to perform a range of operations on network data. Shortest route and connectivity functions are simple forms of network analysis. More complex analyses are often necessary on network data for electrical, gas, and communications applications. The analyses that may be required include simulation of flows in complex networks, load balancing in electrical distribution networks, traffic flow analysis, calculation of pressure loss in gas pipes, and optimization of complex delivery routes with tight constraints of time and load. Network analysis can now be accomplished on a multi-country basis.

Schematics*

Allows you to display any type of network data in various diagrammatic views. Schematic diagrams provide new ways to visualize and analyze relationships and connectivity that may not be otherwise evident. Schematics might be used to investigate utility networks, critical infrastructure, or even spread of disease.

Acronyms

AMI	Amazon Machine Instance
AWS	Amazon Web Services
BLOB	binary large object
CAD	computer-aided design
CAO	chief administrative officer
CASE	computer-aided software engineering
CEO	chief executive officer
CFO	chief financial officer
CIFS	common Internet file services
CIO	chief information officer
COGO	coordinate geometry
CPU	central processing unit
CU	compute unit
DBMS	database management system
DEM	digital elevation model
DPM	displays per minute
ED50	European Datum of 1950
ELB	elastic load balancing
EPC	Enterprise Process Center
GB	gigabyte
GDB	geodatabase
GIO	geographic information officer
GIS	geographic information system
GPS	Global Positioning System
HTTP	hypertext transfer protocol
ICA	independent computing architecture
IPD	information product description
IT	information technology
Kb	kilobit
KB	kilobyte
LAN	local area network
Mb	megabit
MB	megabyte

Mbpd megabits per display
Mbps megabits per second
MDS mosaic dataset
MIDL master input data list
MOU memorandum of understanding
MSD map service description
NAD27 North American Datum of 1927
NAD83 North American Datum of 1983
NFS network file services
NPV net present value
OD origin-destination
OWL Web Ontological Language
PB petabyte
PDIS property development information system
PVB present value of benefits
PVC present value of costs
RAID redundant array of independent disks
RDBMS relational database management system
RDP remote desktop protocol
RFI request for information
RFP request for proposal
RFQ request for qualifications
RIF rule interchange format
SDE spatial database engine
SIMS sewer information management system
SOA service-oriented architecture
SOC server object container
SPEC Standard Performance Evaluation Corporation
TB terabyte
TCP/IP transmission control/Internet protocol
TIN triangulated irregular network
UML Unified Modeling Language
URL uniform resource locator
UTM universal transverse Mercator
VGI volunteered geographic information
VRML virtual reality modeling language
WAN wide area network
WTS Windows Terminal Server

Further reading

Books

Brewer, Cynthia. 2005. *Designing Better Maps: A Guide for GIS Users.* Redlands, Calif.: Esri Press.

DeMers, Michael N. 2008. *Fundamentals of Geographic Information Systems.* 4th ed. New York: John Wiley & Sons, Inc.

Eason, Kenneth. 1989. *Information Technology and Organisational Change.* London: Taylor & Francis.

Fleming, Cory, ed. 2005. *The GIS Guide for Local Government Officials.* Redlands, Calif.: Esri Press.

Foresman, Timothy, ed. 1997. *The History of Geographic Information Systems.* New York: Prentice Hall.

Fu, Pinde, and Jiulin Sun. 2010. *Web GIS: Principles and Applications.* Redlands, Calif.: Esri Press.

Harmon, John E., and Steven J. Anderson. 2003. *The Design and Implementation of Geographic Information Systems.* New York: John Wiley & Sons, Inc.

Kimerling, Jon, Phillip Muehrcke, Juliana Muehrcke, and Aileen Buckley. 2011. *Map Use: Reading, Analysis, Interpretation.* 7th ed. Redlands, Calif.: Esri Press.

Longley, Paul A., Michael F. Goodchild, David J. Maguire, and David W. Rhind. 2010. *Geographic Information Systems and Science.* 3rd ed. New York: John Wiley & Sons, Inc.

Maguire, David, Michael Batty, and Michael Goodchild. 2005. *GIS, Spatial Analysis, and Modeling.* Redlands, Calif.: Esri Press.

Maguire, David, Victoria Kouyoumjian, and Ross Smith. 2008. *The Business Benefits of GIS: An ROI Approach.* Redlands, Calif.: Esri Press.

Mitchell, Andy. 1999. *The ESRI Guide to GIS Analysis, Volume 1: Geographic Patterns and Relationships.* Redlands, Calif.: Esri Press.

Law, Michael, and Amy Collins. 2013. *Getting to Know ArcGIS for Desktop.* 3rd ed. Redlands, Calif.: Esri Press.

O'Sullivan, David, and David Unwin. 2010. *Geographic Information Analysis.* 2nd ed. New York: John Wiley & Sons, Inc.

Peters, Dave. 2012. *Building a GIS: System Architecture Design Strategies for Managers.* 2nd ed. Redlands, Calif.: Esri Press.

Sommers, Rebecca. 2001. *Quick Guide to GIS Implementation and Management.* Park Ridge, Ill.: Urban and Regional Information Systems Association.

Tang, Winnie, and Jan Selwood. 2005. *Spatial Portals: Gateways to Geographic Information.* Redlands, Calif.: Esri Press.

Tomlinson, R. F., and M. A. G. Toomey. 1999. GIS and LIS in Canada. In *Mapping a Northern Land: The Survey of Canada 1947–1994,* ed. Gerald McGrath and Louis Sebert. McGill Queen's University Press.

Wade, Tasha, and Shelly Sommer, eds. 2006. *A to Z GIS: An Illustrated Dictionary of Geographic Information Systems.* Redlands, Calif.: Esri Press.

Zeiler, Michael, and David Arctur. 2004. *Designing Geodatabases: Case Studies in GIS Data Modeling.* Redlands, Calif.: Esri Press.

Journal articles

Buliung, R. N., and P. S. Kanaroglou. 2004. "On Design and Implementation of an Object-relational Spatial Database for Activity/Travel Behaviour Research." *Journal of Geographical Systems* 6 (3): 237–62.

Calkins, Hugh W., and Duane F. Marble. 1987. "The Transition to Automated Production Cartography: Design of the Master Cartographic Database." *The American Cartographer* 14 (2): 105–19.

Haklay, M., and C. Tobón. 2003. "Usability evaluation and PPGIS: Towards a User-Centered Design Approach." *International Journal of Geographical Information Science* 17 (6): 577–92.

Poch, M., J. Comas, et al. 2004. "Designing and Building Real Environmental Decision Support Systems." *Environmental Modelling and Software* 19 (9): 857–73.

Tomlinson, R. F., and Douglas A. Smith. 1991. "Assessing GIS Costs and Benefits: Methodological and Implementation Issues." *International Journal Geographical Information Systems* 6 (3): 247–56.

Wilcox, Darlene L. 2000. "Now What Do We Do? Using Cost-Benefit Analysis for Strategic Planning." *GEOWorld* 13 (2): 42–4.

———. 1990. "Concerning 'The Economic Evaluation of Implementing a GIS.'" *International Journal of Geographical Information Systems* (April–June).

Websites

The following websites include extensive reading on topics of relevance to GIS managers.

Esri. `http://www.esri.com`.

System Design Strategies, an overview of system design philosophy from the Esri perspective, updated regularly by Dave Peters. `http://www.esri.com/systemdesign`.

US Census Bureau. `http://www.census.gov`.

US Geological Survey. `http://usgs.gov`.

Index

D

J

K

L